MW01379965

EVALUATION AND REPAIR OF CONCRETE STRUCTURES

U.S. Army Corps of Engineers

Fredonia Books
Amsterdam, The Netherlands

Evaluation and Repair of Concrete Structures

by
U.S. Army Corps of Engineers

ISBN: 1-4101-0743-4

Reprinted from the 1995 edition

Fredonia Books
Amsterdam, The Netherlands
http://www.fredoniabooks.com

DEPARTMENT OF THE ARMY
U.S. Army Corps of Engineers
Washington, DC 20314-1000

EM 1110-2-2002

CECW-EG

Manual
No. 1110-2-2002

30 June 1995

Engineering and Design
EVALUATION AND REPAIR OF CONCRETE STRUCTURES

Table of Contents

Chapter 1
Introduction

1-1. Purpose

This manual provides guidance on evaluating the condition of the concrete in a structure, relating the condition of the concrete to the underlying cause or causes of that condition, selecting an appropriate repair material and method for any deficiency found, and using the selected materials and methods to repair or rehabilitate the structure. Guidance is also included on maintenance of concrete and on preparation of concrete investigation reports for repair and rehabilitation projects. Considerations for certain specialized types of rehabilitation projects are also given.

1-2. Applicability

This manual is applicable to all HQUSACE elements and USACE commands having civil works responsibilities.

1-3. References

References are listed in Appendix A. Copies of all the references listed should be maintained in their most current versions by districts and divisions having civil works responsibilities. The copies should be kept in a location easily accessible to personnel responsible for concrete condition evaluations and concrete repair projects.

1-4. Definitions and Abbreviations

Terms frequently used in this manual are defined in the Glossary (Appendix B). Also, abbreviations used in this manual are explained in Appendix C.

1-5. Methodology for Repair and Rehabilitation

This manual deals primarily with evaluation and repair of concrete structures; however, a basic understanding of underlying causes of concrete deficiencies is essential to performing meaningful evaluations and successful repairs. If the cause of a deficiency is understood, it is much more likely that the correct repair method will be selected and that, consequently, the repair will be successful. Symptoms or observations of a deficiency must be differentiated from the actual cause of the deficiency, and it is imperative that causes and not symptoms be addressed in repairs. For example, cracking is a symptom of distress that may have a variety of causes. Selection of the correct repair technique for cracking depends upon knowing whether the cracking is caused by repeated freezing and thawing of the concrete, accidental loading, or some other cause. Only after the cause or causes are known can rational decisions be made concerning the selection of a proper method of repair and in determining how to avoid a repetition of the circumstances that led to the problem. The following general procedure should be followed for evaluating the condition and correcting the deficiencies of the concrete in a structure:

a. *Evaluation.* The first step is to evaluate the current condition of the concrete. This evaluation may include a review of design and construction documents, a review of structural instrumentation data, a visual examination, nondestructive testing (NDT), and laboratory analysis of concrete samples. Upon completion of this evaluation step, personnel making the evaluation should have a thorough understanding of the condition of the concrete and may have insights into the causes of any deterioration noted.

b. *Relating observations to causes.* Once the evaluation of a structure has been completed, the visual observations and other supporting data must be related to the mechanism or mechanisms that caused the damage. Since many deficiencies are caused by more than one mechanism, a basic understanding of causes of deterioration of concrete is needed to determine the actual damage-causing mechanism for a particular structure.

c. *Selecting methods and materials.* Once the underlying cause of the damage observed in a structure has been determined, selection of appropriate repair materials and methods should be based on the following considerations:

(1) Prerepair adjustments or modifications required to remedy the cause, such as changing the water drainage pattern, correcting differential foundation subsidence, eliminating causes of cavitation damage, etc.

(2) Constraints such as access to the structure, the operating schedule of the structure, and the weather.

(3) Advantages and disadvantages of making permanent versus temporary repairs.

(4) Available repair materials and methods and the technical feasibility of using them.

(5) Quality of those technically feasible methods and materials to determine the most economically viable to ensure a satisfactory job.

d. Preparation of plans and specifications. The next step in the repair or rehabilitation process is preparation of project plans and specifications. When required by a major rehabilitation project, a Concrete Materials Design Memorandum, in the form of a separate report or a part of the Rehabilitation Evaluation report, should be prepared as outlined in Chapter 9. Existing guide specifications should be used to the maximum extent possible. However, many of the materials and methods described in this manual are not covered in the existing guide specifications. If the materials and methods needed for a particular repair project are not covered in the guide specifications, a detailed specification based upon the guidance given in this manual and upon experience gained from similar projects should be prepared. Since the full extent of concrete damage may not be completely known until concrete removal begins, plans and specifications for repair projects should be prepared with as much flexibility with regard to material quantities as possible. A thorough condition survey, as outlined in Chapter 2, performed as close as possible to the time repair work is executed should help minimize errors in estimated quantities.

e. Execution of the work. The success of a repair or rehabilitation project will depend upon the degree to which the work is executed in conformance with plans and specifications. There is growing evidence, based upon experience gained on a number of projects, that concrete work on repair projects requires much greater attention to good practice than may be necessary for new construction. Because of the importance of the attention to detail and the highly specialized construction techniques required for most repairs, it is important that the design engineer responsible for the investigation of the distress and selection of repair materials and methods be intimately involved in the execution of the work. For example, many repair projects require placing relatively thin overlays, either vertically or horizontally. The potential for cracking in these placements is much greater than it is during placement of concrete in new construction because of the high degree of restraint.

Chapter 2
Evaluation of the Concrete
in Concrete Structures

2-1. Introduction

This chapter presents information on how to conduct an evaluation of the concrete in a concrete structure. As was described in Chapter 1, a thorough and logical evaluation of the current condition of the concrete in a structure is the first step of any repair or rehabilitation project. When the condition of a structure indicates that major repair or rehabilitation is probably necessary, a comprehensive evaluation of the structure should be conducted to determine the scope of the work required. Such an evaluation could include the following: a review of the available design and construction documentation; a review of the operation and maintenance records; a review of the instrumentation data; a visual examination of the condition of the concrete in the structure; an evaluation of the structure by nondestructive testing means; a laboratory evaluation of the condition of concrete specimens recovered from the structure; a stress analysis; and a stability analysis of the entire structure. With the exception of performing stress and stability analyses, each of these general areas is described in detail in this chapter.

2-2. Review of Engineering Data

A thorough review of all of the pertinent data relating to a structure should be accomplished early in the evaluation process. To understand the current condition of the concrete in a structure, it is imperative to consider how design, construction, operation, and maintenance have interacted over the years since the structure was designed and constructed. Sources of engineering data which can yield useful information of this nature include project design memoranda, plans and specifications, construction history reports, as-built drawings, concrete report or concrete records (including materials used, batch plant and field inspection records, and laboratory test data), instrumentation data, operation and maintenance records, and periodic inspection reports. Instrumentation data and monument survey data to detect movement of the structure should be examined.

2-3. Condition Survey

A condition survey involves visual examination of exposed concrete for the purpose of identifying and defining areas of distress. A condition survey will usually include a mapping of the various types of concrete deficiencies that may be found, such as cracking, surface

problems (disintegration and spalling), and joint deterioration. Cracks are usually mapped on fold-out sketches of the monolith surfaces. Mapping must include inspection and delineating of pipe and electrical galleries, filling and emptying culverts (if possible), and other similar openings. Additionally, a condition survey will frequently include core drilling to obtain specimens for laboratory testing and analysis. Stowe and Thornton (1984), American Concrete Institute (ACI) 207.3R, and ACI 364.1R[1] provide additional information on procedures for conducting condition surveys.

a. Visual inspection. A visual inspection of the exposed concrete is the first step in an on-site examination of a structure. The purpose of such an examination is to locate and define areas of distress or deterioration. It is important that the conditions observed be described in unambiguous terms that can later be understood by others who have not inspected the concrete. Terms typically used during a visual inspection are listed by category in Table 2-1. Each of the categories of terms in the table is discussed in detail in the following subparagraphs. Additional descriptions may be found in Appendix B, ACI 116R, and ACI 201.1R.

(1) Construction faults. Typical construction faults that may be found during a visual inspection include bug holes, evidence of cold joints, exposed reinforcing steel, honeycombing, irregular surfaces caused by improperly aligned forms, and a wide variety of surface blemishes and irregularities. These faults are typically the result of poor workmanship or the failure to follow accepted good practice. Various types of construction faults are shown in Figures 2-1 through 2-4.

(2) Cracking. Cracks that occur in concrete may be described in a variety of ways. Some of the more common ways are in terms of surface appearance, depth of cracking, width of cracking, current state of activity, physical state of concrete when cracking occurred, and structural nature of the crack. Various types of cracks based on these general terms are discussed below:

(a) Surface appearance of cracks. The surface appearance of cracks can give the first indication of the cause of cracking. Pattern cracks (Figures 2-5 through 2-7) are rather short cracks, usually uniformly distributed and interconnected, that run in all directions. Pattern cracking indicates restraint of contraction of the surface layer by the backing or inner concrete or possibly an

---

[1] All ACI references are listed with detailed information in Appendix A.

Table 2-1
Terms Associated with Visual Inspection of Concrete

Construction faults	Distortion or movement
Bug holes	Buckling
Cold joints	Curling or warping
Exposed reinforcing steel	Faulting
Honeycombing	Settling
Irregular surface	Tilting
Cracking	
Checking or crazing	Erosion
D-cracking	Abrasion
Diagonal	Cavitation
Hairline	Joint-sealant failure
Longitudinal	Seepage
Map or pattern	Corrosion
Random	Discoloration or staining
Transverse	Exudation
Vertical	Efflorescence
Horizontal	Incrustation
Disintegration	Spalling
Blistering	Popouts
Chalking	Spall
Delamination	
Dusting	
Peeling	
Scaling	
Weathering	

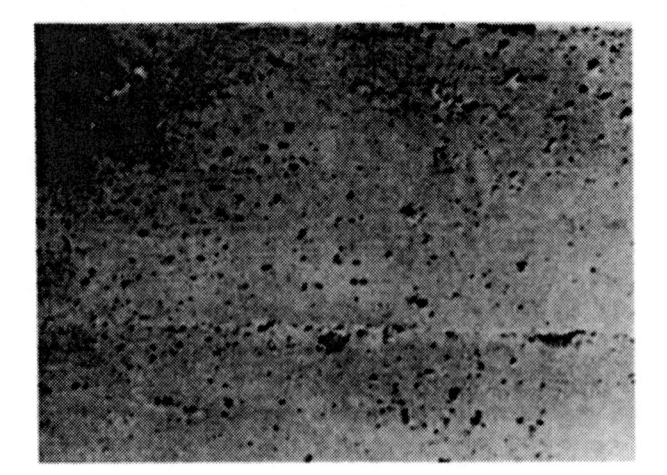

Figure 2-1. Bug holes in a vertical wall

Figure 2-2. Honeycombing and cold joint

Figure 2-3. Cold joint

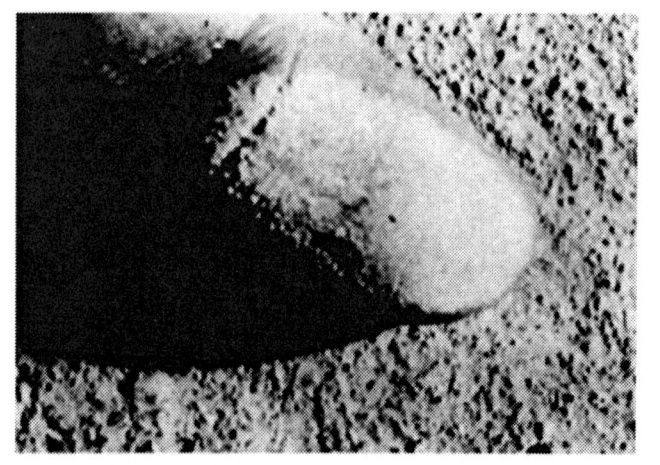

Figure 2-4. Dusting on horizontal finished surface

Figure 2-5. Pattern cracking caused by restrained volume changes

increase of volume in the interior of the concrete. Other terms used to describe pattern cracks are map cracks, crazing, and checking (see Glossary, Appendix B, for definitions). Another type of pattern crack is D-cracking. Figure 2-8 shows typical D-cracking in a concrete pavement. D-cracking usually starts in the lower part of a concrete slab adjacent to joints, where moisture accumulates, and progresses away from the corners of the slab. Individual cracks (Figures 2-9 through 2-11) run in definite directions and may be multiple cracks in parallel at definite intervals. Individual cracks indicate tension in the direction perpendicular to the cracking. Individual cracks are also frequently referred to as isolated cracks. Several terms may be used to describe the direction that an individual or isolated crack runs. These terms include diagonal, longitudinal, transverse, vertical, and horizontal.

(b) Depth of cracking. This category is self-explanatory. The four categories generally used to describe crack depth are surface, shallow, deep, and through.

Figure 2-6. Pattern cracking resulting from alkali-slice reaction

(c) Width of cracking. Three width ranges are used: fine (generally less than 1 mm (0.04 in.)); medium (between 1 and 2 mm (0.04 and 0.08 in.)); and wide (over 2 mm (0.08 in.)) (ACI 201.1R).

Figure 2-7. Pattern cracking caused by alkali-carbonate reaction

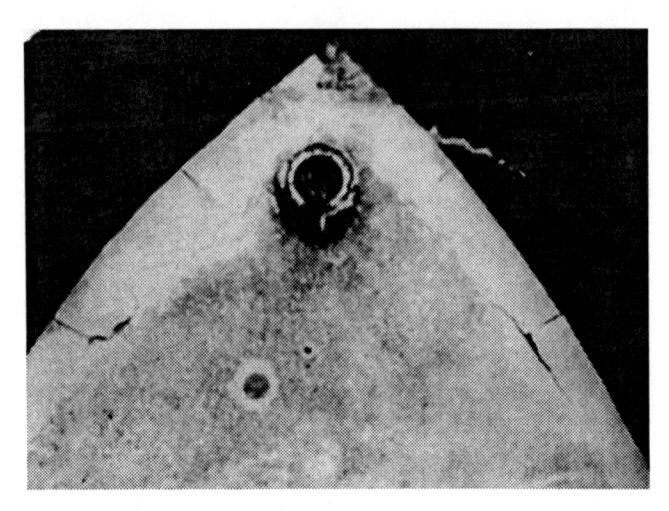

Figure 2-9. Isolated cracks as a result of restraint in the direction perpendicular to the crack

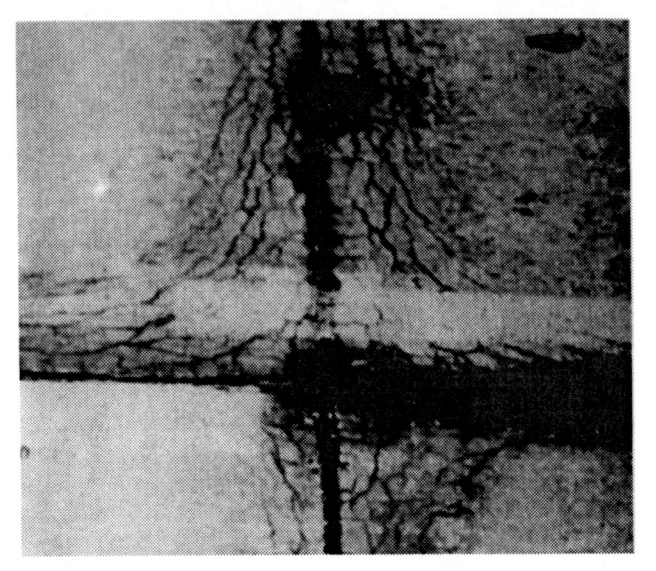

Figure 2-8. D-cracking in a concrete pavement

Figure 2-10. Parallel individual cracking caused by freezing and thawing

(d) Current state of activity. The activity of the crack refers to the presence of the factor causing the cracking. The activity must be taken into account when selecting a repair method. Two categories exist: Active cracks are those for which the mechanism causing the cracking is still at work. If the crack is currently moving, regardless of why the crack formed initially or whether the forces that caused it to form are or are not still at work, it must be considered active. Also, any crack for which an exact cause cannot be determined should be considered active. Dormant cracks are those that are not currently moving or for which the movement is of such magnitude that a repair material will not be affected by the movement.

(e) Physical state of concrete when cracking occurred. Cracks may be categorized according to whether cracking occurred before or after the concrete hardened. This classification is useful to describe cracking that occurs when the concrete is fresh: for example, plastic shrinkage cracks.

(f) Structural nature of the crack. Cracks may also be categorized as structural (caused by excessive live or dead loads) and nonstructural (caused by other means). A structural crack will usually be substantial in width, and the opening may tend to increase as a result of continuous

Figure 2-11. Isolated crack caused by structural overload

loading and creep of the concrete. In general, it can be difficult to determine readily during a visual examination whether a crack is structural or nonstructural. Such a determination will frequently require an analysis by a structural engineer. Any significant isolated crack that is discovered during a visual examination should be referred to a structural engineer and should be considered as possibly structural in nature.

(g) Combinations of descriptions. To describe cracking accurately, it will usually be necessary to use several terms from the various categories listed above. For example: (1) shallow, fine, dormant, pattern cracking that occurred in hardened concrete, (2) shallow, wide, dormant, isolated short cracks that occurred in fresh concrete, (3) through, active, transverse, isolated, diagonal cracks that occurred in hardened concrete.

(3) Disintegration. Disintegration of concrete may be defined as the deterioration of the concrete into small fragments or particles resulting from any cause. Disintegration may be differentiated from spalling by the mass of the particles being removed from the main body of concrete. Disintegration is usually the loss of small particles and individual aggregate particles, while spalling is typically the loss of larger pieces of intact concrete. Disintegration may be the result of a variety of causes including aggressive-water attack, freezing and thawing, chemical attack, and poor construction practices. Disintegration resulting from several different causes is shown in Figures 2-12 through 2-15. As is shown in Table 2-1, a wide variety of terms are used to describe disintegration. These terms are defined in the Glossary. Two of the most frequently used terms to describe particular types of disintegration are scaling and dusting.

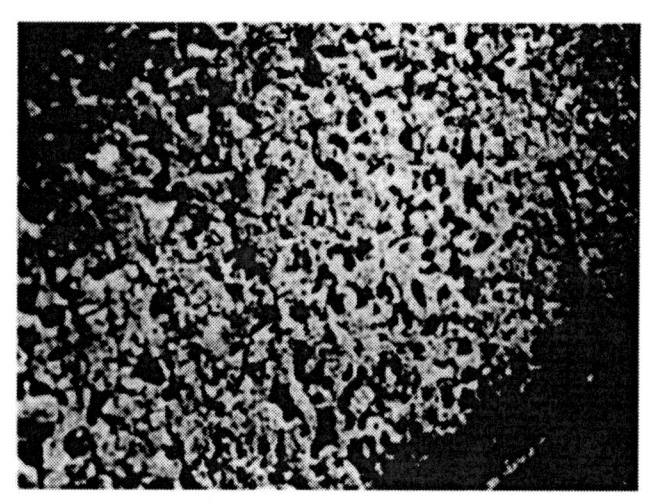

Figure 2-12. Disintegration of concrete caused by exposure to aggressive water

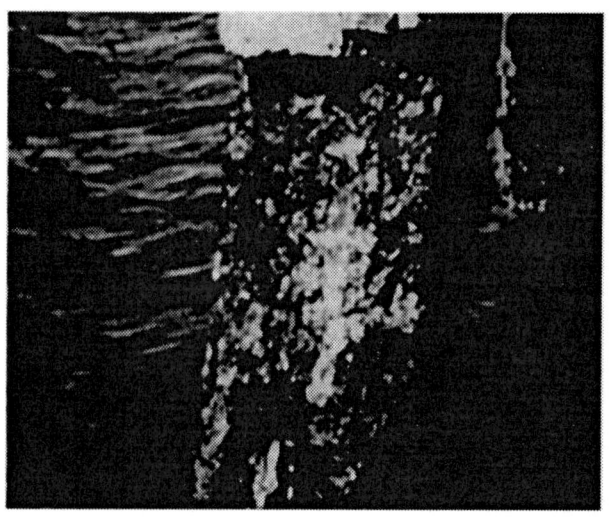

Figure 2-13. Disintegration of concrete caused by exposure to acidic water

(a) Scaling. Scaling is the localized flaking or peeling away of the near-surface portion of the hardened concrete or mortar. Scaling is frequently a symptom of freezing and thawing damage. Degrees of concrete scaling may be defined as follows (ACI 201.1R). Light spalling is loss of surface mortar without exposure of coarse aggregate (Figure 2-16). Medium spalling is loss of surface mortar up to 5 to 10 mm (0.2 to 0.4 in.) in depth and exposure of coarse aggregate (Figure 2-17). Severe spalling is loss of surface mortar 5 to 10 mm (0.2 to 0.4 in.) in depth with some loss of mortar surrounding aggregate particles 10 to 20 mm (0.4 to 0.8 in.) in depth, so that

Figure 2-14. Disintegration of concrete caused by sulfate attack

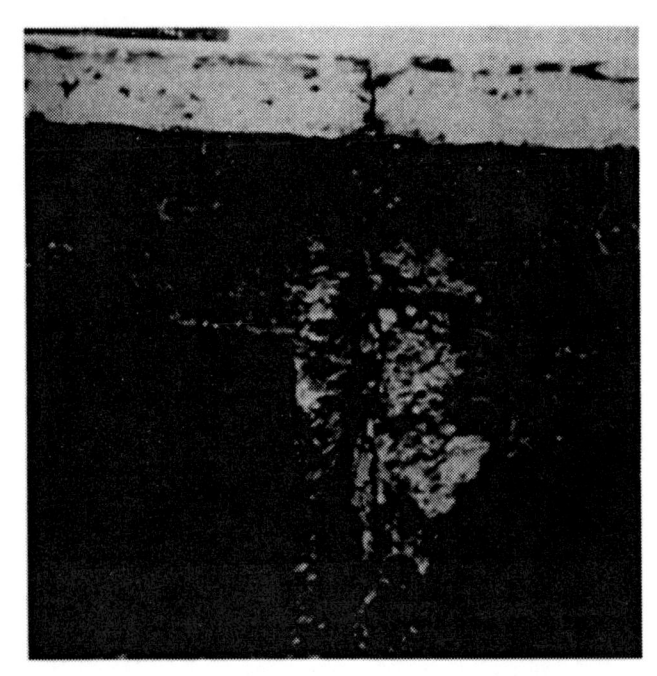

Figure 2-15. Disintegration at a monolith joint as a result of repeated cycles of freezing and thawing and barge impact

Figure 2-16. Light scaling

Figure 2-17. Medium scaling

aggregate is clearly exposed and stands out from the concrete (Figure 2-18). Very severe spalling is loss of coarse aggregate particles as well as surface mortar and surrounding aggregate, generally to a depth greater than 20 mm (0.8 in.) (Figure 2-19).

(b) Dusting. Dusting is the development of a powdered material at the surface of hardened concrete. Dusting will usually be noted on horizontal concrete surfaces

Figure 2-18. Severe scaling

Figure 2-19. Very severe scaling

that receive a great deal of traffic. Typically, dusting is a result of poor construction practice. For example, sprinkling water on a concrete surface during finishing will frequently result in dusting.

(4) Distortion or movement. Distortion or movement, as the terms imply, is simply a change in alignment of the components of a structure. Typical examples would be differential movement between adjacent monoliths or the shifting of supported members on their supports. Review

of historical data such as periodic inspection reports may be helpful in determining when movement first occurred and the apparent rate of movement.

(5) Erosion. Erosion of concrete may be categorized as one of two general types, each of which has a distinct appearance.

(a) Abrasion. Abrasion-erosion damage is caused by repeated rubbing and grinding of debris or equipment on a concrete surface. In hydraulic structures such as stilling basins, abrasion-erosion results from the effects of waterborne gravel, rock, or other debris being circulated over a concrete surface during construction or routine operation. Abrasion-erosion of this type is readily recognized by the smooth, well-worn appearance of the concrete (Figure 2-20).

(b) Cavitation. Cavitation-erosion damage is caused by repeated impact forces caused by collapse of vapor bubbles in rapidly flowing water. The appearance of concrete damaged by cavitation-erosion is generally different from that damaged by abrasion-erosion. Instead of a smooth, worn appearance, the concrete will appear very rough and pitted (Figure 2-21). In severe cases, cavitation-erosion may remove large quantities of concrete and may endanger the structure. Usually, cavitation-erosion occurs as a result of water velocities greater than 12.2 m/sec (40 ft/sec).

(6) Joint sealant failure. Joint sealant materials are used to keep water out of joints and to prevent debris from entering joints and making them ineffective as the concrete expands. Typical failures will be seen as

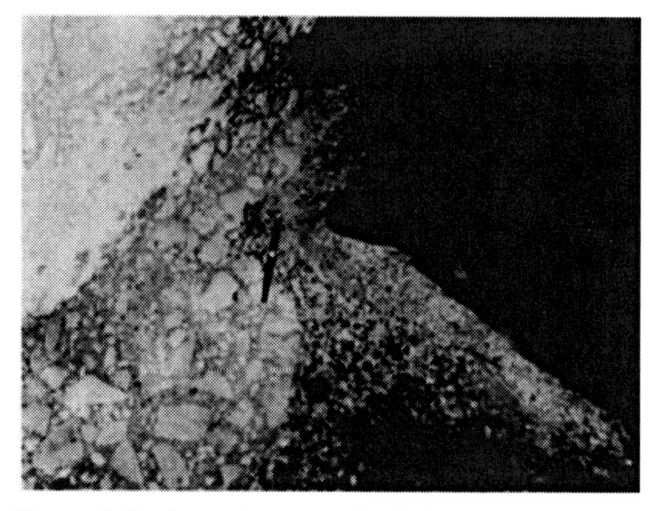

Figure 2-20. Smooth, worn, abraded concrete surface caused by abrasion of waterborne debris

Figure 2-21. Rough, pitted concrete surface caused by cavitation

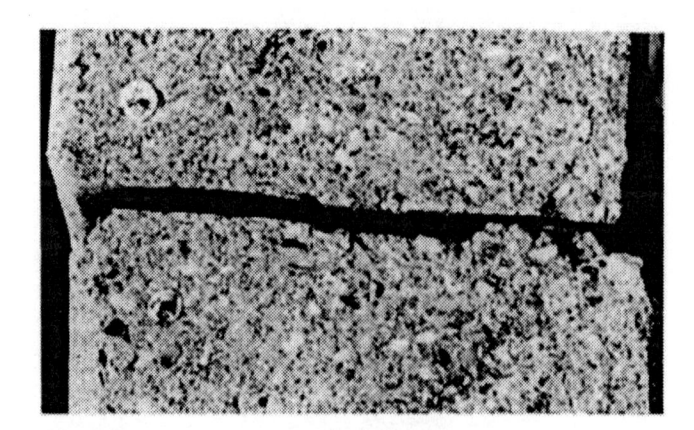

Figure 2-23. Loss of joint sealant

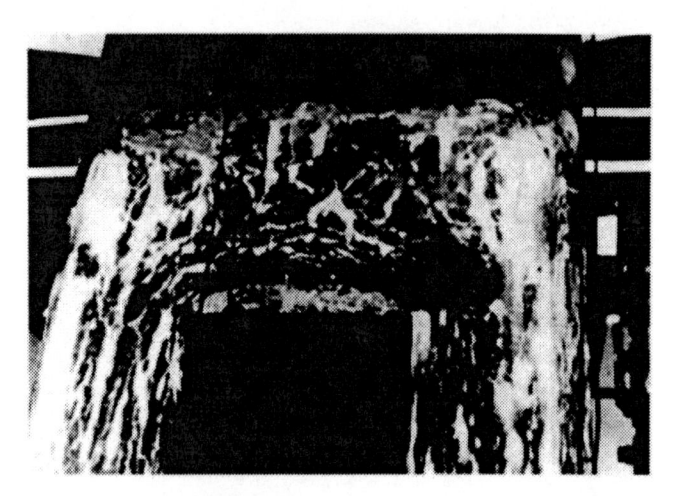

Figure 2-24. Efflorescence

detachment of the sealant material from one or both sides of the joint or complete loss of the sealant material (Figures 2-22 and 2-23).

(7) Seepage. Seepage is defined in ACI 207.3R as "the movement of water or other fluids through pores or interstices." As shown in Table 2-1, the visual evidence of seepage could include, in addition to the presence of water or moisture, evidence of corrosion, discoloration, staining, exudations, efflorescence, and incrustations (Figures 2-24 through 2-28). (For definitions of these terms, see the Glossary, Appendix B). Although occurrences of this nature are quite common around hydraulic structures, they should be included in reports of visual inspections because the underlying cause may be significant. Seepage is another case in which review of historical data may be of benefit to determine whether rates are changing.

(8) Spalling. Spalling is defined as the development of fragments, usually in the shape of flakes, detached from a larger mass. As noted in paragraph 2-3a(3), spalling differs from disintegration in that the material being lost from the mass is concrete and not individual aggregate particles that are lost as the binding matrix disintegrates. The distinction between these two symptoms is important in any attempt to relate symptoms to causes of concrete problems. Spalls can be categorized as follows:

(a) Small spall. Not greater than 20 mm (0.8 in.) in depth nor greater than 150 mm (6 in.) in any dimension (Figure 2-29).

Figure 2-22. Deterioration of joint sealant

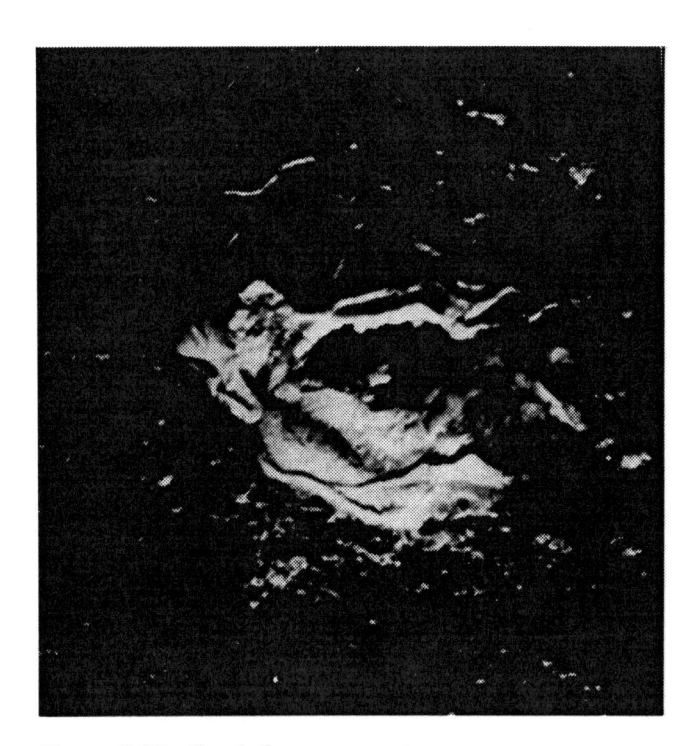

Figure 2-25. Exudation

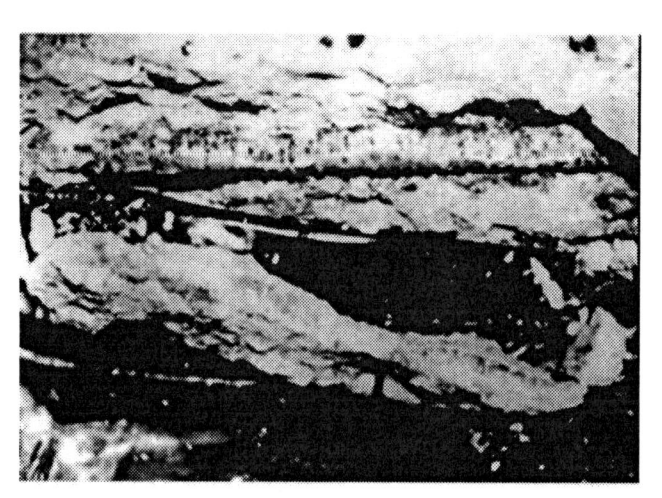

Figure 2-27. Corrosion

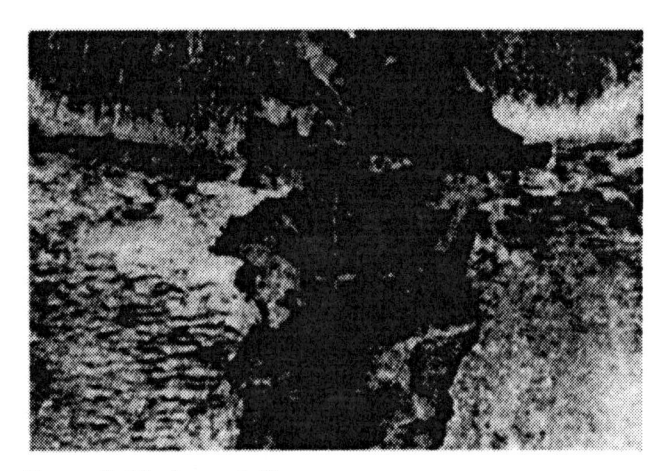

Figure 2-26. Incrustation

Figure 2-28. Water seepage through joint

(b) Large spall. Deeper than 20 mm (0.8 in.) and greater than 150 mm (6 in.) in any dimension (Figure 2-30).

(9) Special cases of spalling. Two special cases of spalling must be noted:

(a) Popouts. Popouts appear as shallow, typically conical depressions in a concrete surface (Figure 2-31). Popouts may be the result of freezing of concrete that contains some unsatisfactory aggregate particles. Instead of general disintegration, popouts are formed as the water

in saturated coarse aggregate particles near the surface freezes, expands, and pushes off the top of the aggregate particle and the superjacent layer of mortar, leaving shallow pits. Chert particles of low specific gravity, limestone containing clay, and shaly materials are well known for this behavior. Popouts are easily recognizable by the shape of the pit remaining in the surface and by a portion

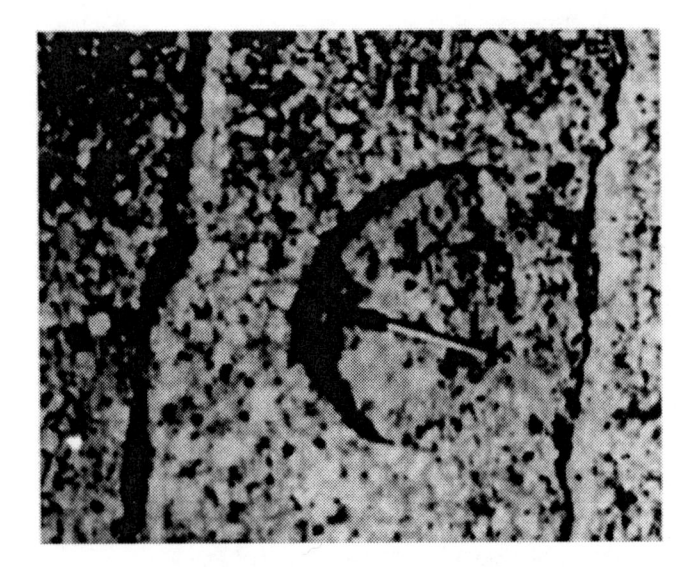

Figure 2-29. Small spall

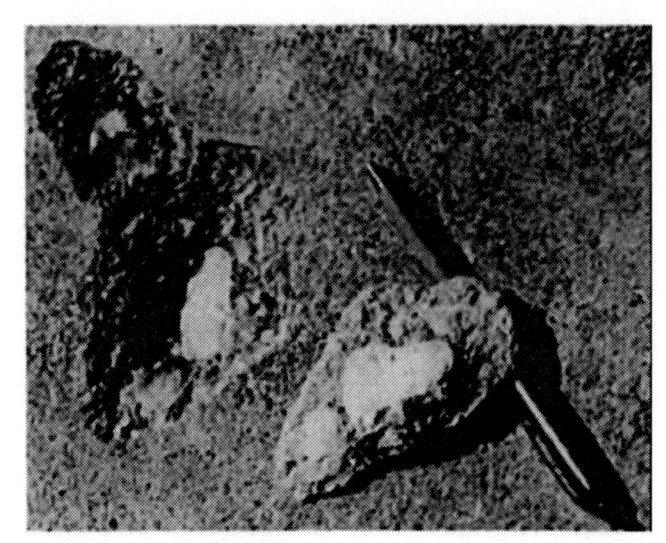

Figure 2-31. Popout

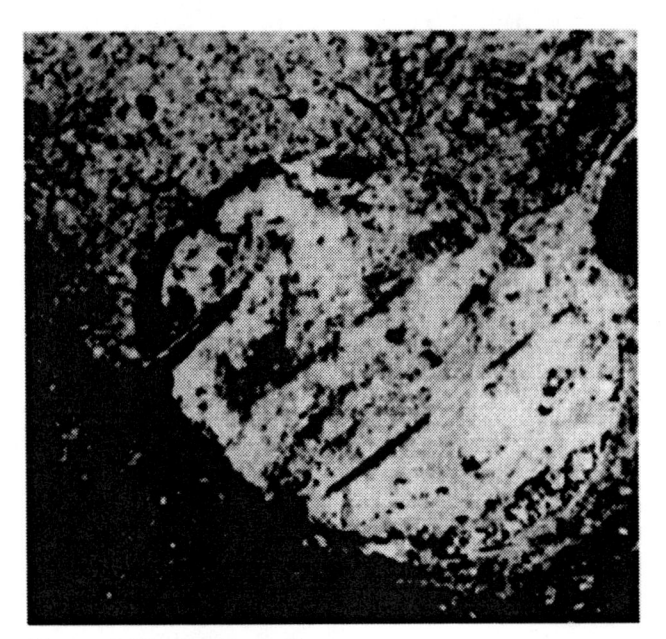

Figure 2-30. Large spall

of the offending aggregate particle usually being visible in the hole (Bach and Isen 1968).

(b) Spalling caused by the corrosion of reinforcement. One of the most frequent causes of spalling is the corrosion of reinforcing steel or other noncorrosion-resistant embedded metal in concrete. During a visual examination of a structure, spalling caused by corrosion of reinforcement is usually an easy symptom to recognize since the corroded metal will be visible along with rust staining,

and the diagnosis will be straightforward. Section 2-3a(10) discusses locating the delamination that occurs before the corrosion progresses to the point that the concrete spalls.

(10) Delamination. Reinforcing steel placed too near the surface or reinforcing steel exposed to chloride ions will corrode. The iron combines with the oxygen in water or air forming rust, or iron oxide, and a corresponding increase in volume up to eight times the original volume. The volume increase results in cracking over the reinforcing steel, followed by delamination along the mat of steel and eventually by spalling. This corrosion sometimes become evident early in the disruptive process when a rectangular pattern of cracking on the concrete surface can be traced to the presence of a reinforcing bar under each crack. Sounding of concrete with a hammer provides a low-cost, accurate method for identifying delaminated areas. Delaminated concrete sounds like a hollow "puck" rather than the "ping" of sound concrete. Boundaries of delaminations can easily be determined by sounding areas surrounding the first "puck" until "pings" are heard.

(a) Hammer-sounding of large areas generally proves to be extremely time consuming. More productive methods are available for sounding horizontal surfaces. Chain dragging accomplishes the same result as hammer-sounding. As the chain is dragged across a concrete surface, a distinctly different sound is heard when it crosses over a delaminated area.

(b) Infrared thermography is a useful method of detecting delaminations in bridge checks. This method is also used for other concrete components exposed to direct sunlight. The method works on the principle that as concrete heats and cools there is substantial thermal gradient within the concrete. Delaminations and other discontinuities interrupt the heat transfer through the concrete. These defects cause a higher surface temperature than that of the surrounding concrete during periods of heating, and a lower surface temperature than that of the surrounding concrete during periods of cooling. The equipment can record and identify areas of delaminations below the surface.

b. Cracking survey. A crack survey is an examination of a concrete structure for the purpose of locating, marking, and identifying cracks and determining the relationship of the cracks with other destructive phenomena (ACI 207.3R). In most cases, cracking is the first symptom of concrete distress. Hence, a cracking survey is significant in evaluating the future serviceability of the structure. The first step in making a crack survey is to locate and mark the cracking and define it by type. The terms for and descriptions of cracks given in Section 2-3 should be used to describe any cracking that is found.

(1) Crack widths can be estimated using a clear comparator card having lines of specified width marked on the card. Crack widths can be measured to an accuracy of about 0.025 mm (0.001 in.) with a crack comparator, a hand-held microscope with a scale on the lens closest to the surface being viewed (Figure 2-32). Crack movement can be monitored with a crack measuring device. The crack monitor shown in Figure 2-33 gives a direct reading of crack displacement and rotation. It is important to make an initial reading when the monitor is attached because the monitor will not necessarily read zero after installation. If more accurate and detailed time histories

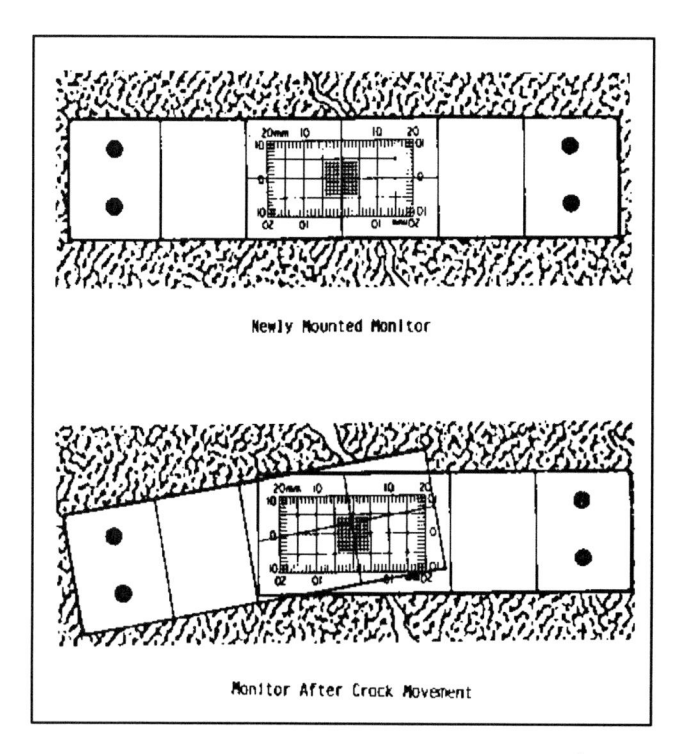

Figure 2-33. Crack monitor

are desired, a wide range of transducers and devices are available (EM 1110-2-4300).

(2) If possible, the crack depth should be determined by observation of edges or insertion of a fine wire or feeler gauge; however, in most situations, the actual depth may be indeterminable without drilling or using other detection techniques such as the pulse-velocity method described in Section 2-6c.

(3) Conditions which may be associated with cracking either over portions of the length or for the entire length should be noted. These conditions may include seepage through the cracks, deposits from leaching or other sources, spalling of edges, differential movement (offsets), etc. Chemical analyses of the seepage water and the deposits may be desirable.

(4) It may be worthwhile to repeat the survey under various loading conditions when change in crack width is suspected. Furthermore, tapping of surfaces with a hammer may detect shallow cracking beneath and parallel to the surface. A hollow sound generally indicates that such cracking is likely even though it cannot be seen. See Section 2-3a(10) for additional discussion on sounding to detect delamination.

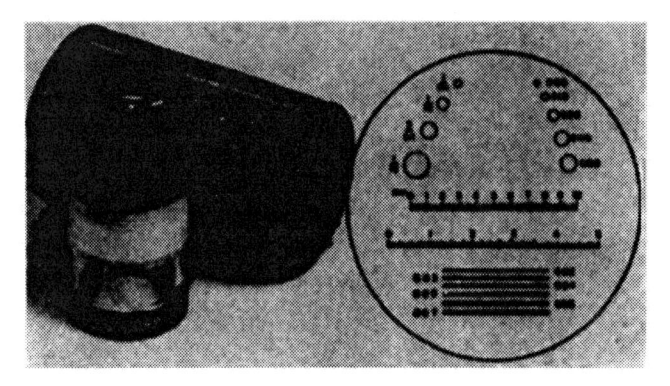

Figure 2-32. Comparator for measuring crack widths

c. Surface mapping.

(1) Surface mapping is a parallel procedure to a cracking survey in which deterioration of the surface concrete is located and described. Surface mapping may be accomplished by use of detailed drawings, photographs, movies, or video tapes. Items most often identified and mapped include: cracking, spalling, scaling, popouts, honeycombing, exudation, distortion, unusual discoloration, erosion, cavitation, seepage, conditions of joints and joint materials, corrosion of reinforcement (if exposed), and soundness of surface concrete. A list of items recommended for use in a surface mapping by hand is as follows (ACI 207.3R):

(a) Structure drawings, if available.

(b) Clipboard and paper or field book.

(c) Tape measure, 15 to 30 m (50 to 100 ft).

(d) Ruler graduated in 1/16 in. or 1 mm.

(e) Feeler gauge.

(f) Pocket comparator or hand microscope.

(g) Knife.

(h) Hammer, 1 kg (2 lb).

(i) Fine wire (not too flexible).

(j) String.

(k) Flashlight or lantern.

(l) Camera with flash and assortment of lenses.

(m) Assortment of film, color and high speed.

(2) Mapping should begin at one end of the structure and proceed in a systematic manner until all surfaces are mapped. Both external and internal surfaces should be mapped if access is possible. Use of three-dimensional (3-D) isometric drawings showing offsets or distortion of structural features is occasionally desirable. Areas of significant distress should be photographed for later reference. A familiar object or scale should be placed in the area to show the relative size of the feature being photographed. It is important to describe each condition mapped in clear, concise detail and to avoid generalizations unless reference is being made to conditions previously detailed in other areas. Profiles are advantageous for showing the depth of erosion.

d. Joint survey. A joint survey is a visual inspection of the joints in a structure to determine their condition. Expansion, contraction, and construction joints should be located and described and their existing condition noted. Opened or displaced joints (surface offsets) should be checked for movement if appropriate; various loading conditions should be considered when measurements of joints are taken. All joints should be checked for defects; for example, spalling or D-cracking, chemical attack, evidence of seepage, emission of solids, etc. Conditions of joint filler, if present, should be examined.

e. Core drilling. Core drilling to recover concrete for laboratory analysis or testing is the best method of obtaining information on the condition of concrete within a structure. However, since core drilling is expensive, it should only be considered when sampling and testing of interior concrete is deemed necessary.

(1) The presence of abnormal conditions of the concrete at exposed surfaces may suggest questionable quality or a change in the physical or chemical properties of the concrete. These conditions may include scaling, leaching, and pattern cracking. When such observations are made, core drilling to examine and sample the hardened concrete may be necessary.

(2) Depth of cores will vary depending upon intended use and type of structure. The minimum depth of sampling concrete in massive structures should be 2 ft in accordance with Concrete Research Division (CRD)-C 26[1] and American Society for Testing and Materials (ASTM) C 823[2]. The core samples should be sufficient in number and size to permit appropriate laboratory examination and testing. For compressive strength, static or dynamic modulus of elasticity, the diameter of the core should not be less than three times the nominal maximum size of aggregate. For 150-mm (6-in.) maximum size aggregate concrete, 200- or 250-mm (8- or 10-in.)-diam cores are generally drilled because of cost, handling, and laboratory testing machine capabilities. Warning should be given against taking NX size 54-mm

---

[1] All CRD-C designations are from U.S. Army Engineer Waterways Experiment Station (USAEWES). 1949 (Aug). *Handbook for Concrete and Cement*, with quarterly supplements, Vicksburg, MS.

[2] All ASTM test methods cited are from the *Annual Book of ASTM Standards* (ASTM Annual).

(2-1/8-in.)-diam cores in concrete. When 50- to 150-mm (2- to 6-in.) maximum size aggregate concrete is cored, an NX size core will generally be recovered in short pieces or broken core. The reason for breakage is that there is simply little mortar bonding the concrete across the diameter of the core. Thus, the drilling action can easily break the core. When drilling in poor-quality concrete with any size core barrel, the material generally comes out as rubble.

(3) Core samples must be properly identified and oriented with permanent markings on the material itself when feasible. Location of borings must be accurately described and marked on photographs or drawings. Cores should be logged by methods similar to those used for geological subsurface exploration. Logs should show, in addition to general information on the hole, conditions at the surface, depth of obvious deterioration, fractures and conditions of fractured surfaces, unusual deposits, coloring or staining, distribution and size of voids, locations of observed construction joints, and contact with the foundation or other surface (ACI 207.3R). The concrete should be wrapped and sealed as may be appropriate to preserve the moisture content representative of the structure at the time of sampling and should be packed so as to be properly protected from freezing or damage in transit or storage, especially if the concrete is very weak. Figure 2-34 illustrates a typical log for a concrete core recovered during a condition survey.

(4) When drill hole coring is not practical or core recovery is poor, a viewing system such as a borehole camera, bore hole television, or borehole televiewer may be used for evaluating the interior concrete conditions. A description and information on the availability of these borehole viewing systems can be found in EP 1110-1-10. Evaluation of distress in massive concrete structures may be desirable to determine in situ stress conditions. ACI 207.3R is an excellent guide to determining existing stress conditions in the structure.

2-4. Underwater Inspection

A variety of procedures and equipment for conducting underwater surveys are available (Popovics and McDonald 1989). Included are several nondestructive techniques which can be used in dark or turbid conditions that preclude visual inspection. Some techniques originally developed for other purposes have been adapted for application in underwater inspections. Prior to an underwater survey, it is sometimes necessary for the surface of the structure to be cleaned. A number of procedures and devices for underwater cleaning of civil works structures are described by Keeney (1987).

a. Visual inspection by divers. Underwater surveys by divers are usually either scuba or surface-supplied diving operations. Basic scuba diving equipment is an oxygen tank, typically weighing about 34 kg (75 lb) which is carried by the diver. Surface-supplied diving, where the air supply is provided from the surface or shore, is a more elaborate operation in terms of equipment, safety concerns, diver skills, etc., especially when the diver approaches maximum allowable depths. Diver equipment for surface-supplied diving includes air compressors, helmets, weighted shoes, air supply lines, breastplates, etc., which can weigh as much as 90 kg (200 lb). The free-swimming scuba diver has more flexibility and maneuverability than the surface-supplied diver. However, he cannot dive as deep or stay underwater as long as a surface-supplied diver.

(1) Advantages. Underwater inspections performed by divers offer a number of advantages: they are (a) applicable to a wide variety of structures; (b) flexible inspection procedures; (c) simple (especially the scuba diver in shallow-water applications); and in most cases, (d) relatively inexpensive. Also, a variety of commercially available instruments for testing concrete above water have been modified for underwater use by divers. These instruments include a rebound hammer to provide data on concrete surface hardness, a magnetic reinforcing steel locator to locate and measure the amount of concrete cover over the reinforcement, and direct and indirect ultrasonic pulse-velocity systems which can be used to determine the general condition of concrete based on sound velocity measurements (Smith 1987).

(2) Limitations. Limitations on diver inspections include the regulations (Engineer Manual 385-1-1) that restrict the allowable depths and durations of dives and the number of repeat dives in a given period. Also, in turbid water a diver's visibility may be reduced to only a few inches, or in extreme cases, a diver may be limited to a tactile inspection. Also, cold climates tend to reduce the diver's ability to perform at normal levels. In any case, a diver's visual, auditory, tactile, and spatial perceptions are different underwater from what they are in air. Therefore, he is susceptible to making errors in observations and recording data.

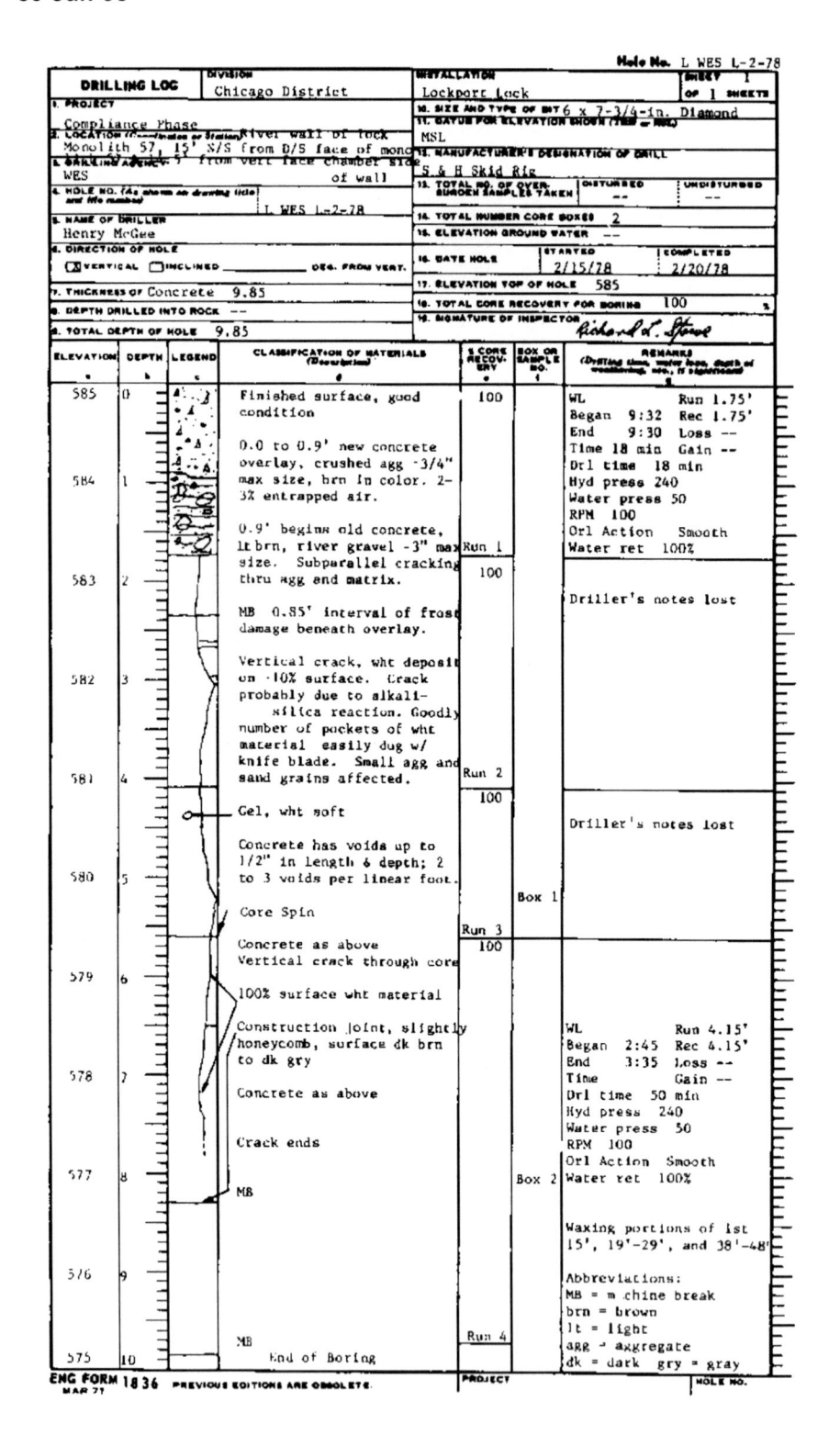

Figure 2-34. Typical information included on a drill log for concrete core

b. Manned and unmanned underwater vehicles.

(1) Underwater vehicles can be thought of as platformed, underwater camera systems with manipulator and propulsion systems. They consist of a video unit, a power source for propulsion, vehicle controllers (referred to as "joysticks"), and display monitor. Available accessories which allow the vehicles to be more functional include angle lens, lighting components, instrumentations for analyses, attachments for grasping, and a variety of other capabilities.

(2) There are five categories of manned underwater vehicles: (a) untethered, (b) tethered, (c) diver lockout, (d) observation/work bells, and (e) atmospheric diving suits. All are operated by a person inside, have viewports, are dry inside the pressure hull(s), and have some degree of mobility.

(3) There are six types of unmanned underwater vehicles: (a) tethered, free swimming, (b) towed, midwater, (c) towed, bottom-reliant, (d) bottom-crawling, (e) structurally reliant, and (f) untethered (Busby Associates, Inc. 1987). These remotely operated vehicles (ROV's) are primarily distinguished by their power source. All include a TV camera to provide real-time or slow-scan viewing, and all have some degree of mobility. They are controlled from the surface via operator-observed video systems. Joysticks are used to control propulsion and manipulation of the ROV and accessory equipment. Exceptions are the untethered types of ROV's which are self-propelled and operated without any connection to the surface. Most ROV's are capable of accommodating various attachments for grasping, cleaning, and performing other inspection chores. Specially designed ROV's can accommodate and operate nondestructive testing equipment.

(4) Underwater vehicles can compensate for the limitations inherent in diver systems because they can function at extreme depths, remain underwater for long durations, and repeatedly perform the same mission without sacrifice in quality. Also, they can be operated in environments where water temperatures, currents, and tidal conditions preclude the use of divers.

(5) Manned underwater vehicles are usually large and bulky systems which require significant operational support. Therefore, they are used less frequently than the smaller unmanned ROV's. Although the dependability of ROV's has steadily increased, some limitations remain. Most ROV systems provide two-dimensional (2-D) views only and, therefore, may not project the full extent of any

defects. Murky water limits the effectiveness of ROV systems. In some situations, it may be difficult to determine the exact orientation or position of the ROV, thus impeding accurate identification of an area being observed (U.S. Dept. of Transportation 1989). Also, ROV's do not possess the maneuverability offered by divers. As a result, controlling the ROV in "tight" areas and in swift currents is difficult and can result in entanglement of the umbilical (REMR Technical Note CS-ES-2.6 (USAEWES 1985a)).

(6) Underwater vehicles are being increasingly accepted as a viable means to effectively perform underwater surveys in practically all instances where traditional diver systems are normally used. Manned underwater vehicles have been used in the inspection of stilling basins, in direct support of divers, and in support of personnel maintaining and repairing wellheads. Applications of ROV's include inspection of dams, breakwaters, jetties, concrete platforms, pipelines, sewers, mine shafts, ship hulls, etc. (Busby Associates, Inc. 1987). They have also been used in leak detection and structure cleaning.

c. Photography systems.

(1) Photography systems used in underwater inspection include still-photography equipment, video recording systems, video imaging systems, and any accessories.

(2) Still-photographic equipment includes cameras, film, and lighting. Most above-water cameras ranging from the "instamatic" type to sophisticated 35-mm cameras can be used underwater in waterproof cases (U.S. Dept. of Transportation 1989). There are also waterproof 35-mm cameras designed specifically for underwater photography (REMR Technical Note CS-ES-3.2 (USAEWES 1985b)). These cameras usually include specially equipped lens and electronic flashes to compensate for the underwater environment. Most film, color and black and white, can be used in underwater photography if ample lighting is provided. High-speed film that compensates for inherent difficulties in underwater photography is available.

(3) Underwater video equipment has improved dramatically in recent years (REMR Technical Note CS-ES-2.6 (USAEWES 1985a)). Video cameras can be used with an umbilical cable to the surface for real-time viewing on a monitor or for recording. Compact camera-recorder systems in waterproof housings can be used with or without the umbilical to the surface. These video systems can be configured to provide on-screen titles and

clock, as well as narration by a diver and surface observer.

(4) Video systems can provide pictorial representations of existing conditions, transmit visual data to topside personnel for analysis and interpretation, and provide a permanent record of the inspection process. Visual recordings can be used to monitor the performance of a structure with time. Additionally, video systems can penetrate turbid areas where the human eye cannot see. Video systems are typically used concurrently with divers and underwater vehicles.

d. High-resolution acoustic mapping system.

(1) Erosion and faulting of submerged surfaces have always been difficult to accurately map. To see into depressions and close to vertical surfaces requires a narrow beam. Also, there is a need to record exactly where a mapping system is located at any instant so that defects may be precisely located and continuity maintained in repeat surveys. These capabilities are provided by the high-resolution acoustic mapping system developed through a joint research and development effort between the U.S. Army Corps of Engineers and the U.S. Bureau of Reclamation (Thornton 1985 and Thornton and Alexander 1987).

(2) The system can be broken into three main components: the acoustic subsystem, a positioning subsystem, and a compute-and-record subsystem. The acoustic subsystem consists of a boat-mounted transducer array and the signal processing electronics. During a survey, each transducer generates acoustic signals which are reflected from the bottom surface and received at the transducer array. The time of flight for the acoustic signal from the transducer to the bottom surface and back is output to a computer. The computer calculates the elevation of the bottom surface from this information, and the basic data are recorded on magnetic disks.

(3) The lateral positioning subsystem consists of a sonic transmitter on the boat and two or more transponders in the water at known or surveyed locations. As each transponder receives the sonic pulse from the transmitter, it radios the time of detection of the survey boat. The position of the boat is calculated from this information and displayed by an onboard computer. The network can be easily reestablished, making it possible to return the survey boat and transducer array bar to a specific location.

(4) The compute-and-record subsystem provides for computer-controlled operation of the system and for processing, display, and storage of data. Survey results are in the form of real-time strip charts showing the absolute relief for each run, 3-D surface relief plots showing composite data from all the survey runs in a given area, contour maps selected areas, and printouts of the individual data points.

(5) The high-resolution acoustic mapping system is designed to operate in water depths of 1.5 m to 12 m (5 to 40 ft) and produce accuracies of ± 50 mm (2 in.) vertically and ± 0.3 m (1 ft) laterally. The major limitation of the system is that it can be used only in relatively calm water. Wave action causing a roll angle of more than 5 deg will automatically shut down the system.

(6) To date, the primary application of this system has been in rapid and accurate surveying of erosion damage in stilling basins. The system has been successfully used at a number of BuRec and Corps of Engineers (CE) dams including Folsom, Pine Flat, Ice Harbor, Locks and Dams 25 and 26 (Miss. River), Lookout Point, and Dexter.

e. Side-scan sonar.

(1) The side-scan sonar, which evolved from the echo sounding depth finders developed during World War II, basically consists of a pair of transducers mounted in a waterproof housing referred to as a "fish," a graphic chart-recorder set up for signal transmission and processing, and tow cable which connects the "fish" and recorder. The system directs sound waves at a target surface. The reflected signals are received by the transducers and transmitted to the chart-recorder as plotted images. The recorded image, called a sonograph, is characterized by various shades of darkened areas, or shadows, on the chart. Characteristics of the reflecting surface are indicated by the intensity of the reflected signals. Steel will reflect a more intense signal and produce a darker shaded area than wood, and gravel will reflect a more intense signal than sand. Acoustic shadows, shades of white, are projected directly behind the reflecting surface. The width of these shadows and the position of the object relative to the towfish are used to calculate the height of the object (Morang 1987).

(2) Electronic advances in the side-scan sonar have broadened its potential applications to include underwater surveying. In the normal position, the system looks at

vertical surfaces. However, it can be configured to look downward at horizontal surfaces in a manner similar to that of the high-resolution acoustic mapping system. The side-scan sonar is known for its photograph-like image. Current commercial side-scan sonar systems are available with microprocessors and advanced electronic features (built in or as accessory components) to print sonographs corrected for slant-range and true bottom distances (Clausner and Pope 1988).

(3) Side-scan sonar has proven useful in surveys of breakwaters, jetties, groins, port structures, and inland waterway facilities such as lock and dams. It has proven especially effective in examing the toe portion of rubble structures for scour and displacement of armor units (Kucharski and Clausner 1990). The ability of sonar to penetrate waters too turbid or dangerous for visual or optical inspection makes it the only effective means of inspecting many coastal structures.

f. Radar.

(1) Radar and acoustics work in a similar manner, except radar uses an electromagnetic signal which travels very fast compared to the relatively slow mechanical wave used in acoustics. In both cases, the time of arrival (TOA) is measured and a predetermined calibration velocity is used to calculate the depth of the reflecting interface. The two main factors that influence radar signals are electrical conductivity and dielectric constant of the material (Alongi, Cantor, Kneeter, and Alongi 1982 and Morey 1974). The conductivity controls the loss of energy and, therefore, the penetration depth. The dielectric constant determines the propagation velocity.

(2) The resistivity (reciprocal of conductivity) of concrete structures varies considerably in the dry, and the presence of water further complicates the measurement. Therefore, those who have a need for this type of underwater survey should contact one or more of the sources referenced for assistance in determining the proper measurement system for a given application.

g. Ultrasonic pulse velocity.

(1) Ultrasonic pulse velocity provides a nondestructive method for evaluating structures by measuring the time of travel of acoustic pulses of energy through a material of known thickness (Thornton and Alexander 1987). Piezoelectric transducers, housed in metal casings and excited by high-impulse voltages, transmit and receive the acoustic pulses. An oscilloscope configured in the

system measures time and displays the acoustic waves. Dividing the length of the travel path by the travel time yields the pulse velocity, which is proportioned to the dynamic modulus of elasticity of the material. Velocity measurements through materials of good quality usually result in high velocities and signal strengths, while materials of poor quality usually exhibit decreased velocities and weak signals. For example, good quality, continuous concrete produces velocities in the range of 3,700 to 4,600 mps (12,000 to 15,000 fps); poor quality or deteriorated concrete, 2,400 to 3,000 mps (8,000 to 10,000 fps).

(2) The pulse-velocity method has provided reliable in situ delineations of the extent and severity of cracks, areas of deterioration, and general assessments of the condition of concrete structures for many years. The equipment can penetrate approximately 91 m (300 ft) of continuous concrete with the aid of amplifiers, is easily portable, and has a high data acquisition-to-cost ratio. Although most applications of the pulse-velocity method have been under dry conditions, the transducers can be waterproofed for underwater surveys.

h. Ultrasonic pulse-echo system.

(1) A new improved prototype ultrasonic pulse-echo (UPE) system for evaluating concrete has been developed by the U.S. Army Engineer Waterways Experiment Station (CEWES). The new system (Alexander and Thornton 1988 and Thornton and Alexander 1987) uses piezoelectric crystals to generate and detect signals and the accurate time base of an oscilloscope to measure the TOA of a longitudinal ultrasonic pulse in concrete.

(2) Tests have shown that the system is capable of delineating sound concrete, concrete of questionable quality, and deteriorated concrete, as well as delaminations, voids, reinforcing steel, and other objects within concrete. Also, the system can be used to determine the thickness of a concrete section in which only one surface is accessible. The system will work on vertical or horizontal surfaces. However, the present system is limited to a thickness of about 0.5 m (1.5 ft). For maximum use of this system, the operator should have had considerable experience using the system and interpreting the results.

(3) The system, which was originally developed to operate in a dry environment, was adapted for use in water to determine the condition of a reinforced concrete sea wall at a large marina (Thornton and Alexander 1988).

i. Sonic pulse-echo technique for piles.

(1) A sonic pulse-echo technique for determining the length of concrete and timber piles in dry soil or underwater has been developed at WES (Alexander 1980). Sonic energy is introduced into the accessible end of the pile with a hammer. If the pulse length generated by the hammer is less than round-trip echo time in the pile, then the TOA can be measured with the accurate time base of an oscilloscope. With a digital oscilloscope, the signal can be recorded on magnetic disc and the signal entered into the computer for added signal processing. If the length, mass, and hardness of the head of the hammer is such that the hammer generates energy in the frequency range that corresponds to the longitudinal resonant frequency of the pile, then the frequency can be measured with a spectrum analyzer.

(2) In addition to determining pile lengths to depths of tens of feet, this system can also detect breaks in a pile. Because the surrounding soil dissipates the energy from the hammer, the length-to-diameter ratio of the pile should be greater than 5 and less than 30. To date, work has been limited only to those applications where the impact end of the pile was above water.

2-5. Laboratory Investigations

Once samples of concrete have been obtained, whether by coring or other means, they should be examined in a qualified laboratory. In general, the examination should include petrographic, chemical, or physical tests. Each of these examinations is described in this paragraph.

a. Petrographic examination. Petrographic examination is the application of petrography, a branch of geology concerned with the description and classification of rocks, to the examination of hardened concrete, a synthetic sedimentary rock. Petrographic examination may include visual inspection of the samples, visual inspection at various levels of magnification using appropriate microscopes, X-ray diffraction analysis, differential thermal analysis, X-ray emission techniques, and thin section analysis. Petrographic techniques may be expected to provide information on the following (ACI 207.3R): (1) condition of the aggregate; (2) pronounced cement-aggregate reactions; (3) deterioration of aggregate particles in place; (4) denseness of cement paste; (5) homogeneity of the concrete; (6) occurrence of settlement and bleeding of fresh concrete; (7) depth and extent of carbonation; (8) occurrence and distribution of fractures; (9) characteristics and distribution of voids; and (10) presence of contaminating substances. Petrographic

examination of hardened concrete should be performed in accordance with ASTM C 856 (CRD-C 57) by a person qualified by education and experience so that proper interpretation of test results can be made.

b. Chemical analysis. Chemical analysis of hardened concrete or of selected portions (paste, mortar, aggregate, reaction products, etc.) may be used to estimate the cement content, original water-cement ratio, and the presence and amount of chloride and other admixtures.

c. Physical analysis. The following physical and mechanical tests are generally performed on concrete cores:

(1) Density.

(2) Compressive strength.

(3) Modulus of elasticity.

(4) Poisson's ratio.

(5) Pulse velocity.

(6) Direct shear strength of concrete bonded to foundation rock.

(7) Friction sliding of concrete on foundation rock.

(8) Resistance of concrete to deterioration caused by freezing and thawing.

(9) Air content and parameters of the air-void system.

Testing core samples for compressive strength and tensile strength should follow the method specified in ASTM C 42 (CRD-C 27).

2-6. Nondestructive Testing

The purpose of NDT is to determine the various relative properties of concrete such as strength, modulus of elasticity, homogeneity, and integrity, as well as conditions of strain and stress, without damaging the structure. Selection of the most applicable method or methods of testing will require good judgment based on the information needed, size and nature of the project, site conditions and risk to the structure (ACI 207.3R). Proper utilization of NDT requires a "toolbox" of techniques and someone with the expertise to know the proper tool to use in the various circumstances. In this paragraph, the commonly used nondestructive testing techniques for evaluating in situ concrete will be discussed. Malhotra (1976), Thornton and Alexander (1987), and Alexander (1993) provide additional information on NDT techniques. Also, recent advances in nondestructive testing of concrete are summarized by Carino (1992). Test methods are classified into those used to assess in-place strength and those used to locate hidden defects. In the first category, recent developments are presented on the pullout test, the break-off test, the torque test, the pulloff test, and the maturity method. In the second category, a review is presented of infrared thermography, ground penetrating radar, and several methods based upon stress wave propagation. The principles of the methods, their advantages, and their inherent limitations are discussed. Where appropriate, requirements of relevant ASTM standards are discussed.

a. Rebound number (hammer).

(1) Description.

(a) The rebound number is obtained by the use of a hammer that consists of a steel mass and a tension spring in a tubular frame (Figure 2-35). When the plunger of the hammer is pushed against the surface of the concrete, the steel mass is retracted and the spring is compressed. When the mass is completely retracted, the spring is automatically released and the mass is driven against the plunger, which impacts the concrete and rebounds. The rebound distance is indicated by a pointer on a scale that is usually graduated from 0 to 100. The rebound readings are termed R-values. Determination of R-values is outlined in the manual supplied by the hammer manufacturer.

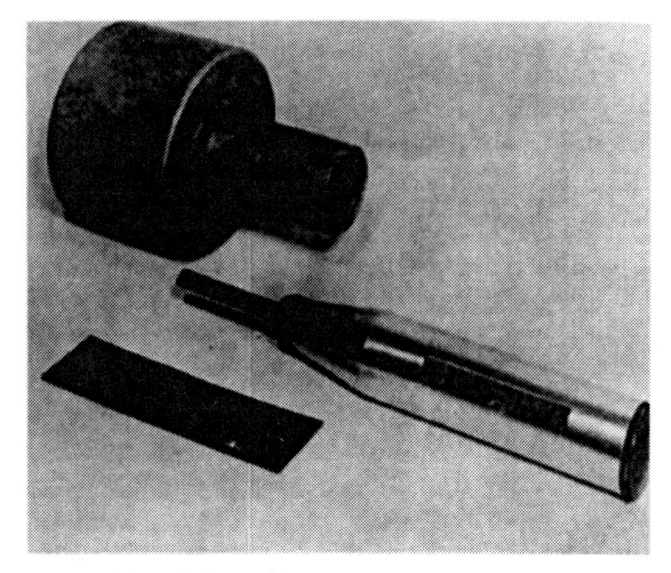

Figure 2-35. Rebound hammer

R-values indicate the coefficient of restitution of the concrete; the values increase with the "strength" of the concrete.

(b) Most hammers come with a calibration chart, showing a purported relationship between compressive strength of concrete and rebound readings. However, rather than placing confidence in such a chart, users should develop their own relations for each concrete mixture and each rebound hammer.

(2) Applications. Rebound numbers may be used to estimate the uniformity and quality of concrete. The test method is covered in ASTM C 805 (CRD-C 22).

(3) Advantages. The rebound hammer is a simple and quick method for NDT of concrete in place. The equipment is inexpensive and can be operated by field personnel with a limited amount of instruction. The rebound hammer is very useful in assessing the general quality of concrete and for locating areas of poor quality concrete. A large number of measurements can be rapidly taken so that large exposed areas of concrete can be mapped within a few hours.

(4) Limitations. The rebound method is a rather imprecise test and does not provide a reliable prediction of the strength of concrete. Rebound measurements on in situ concrete are affected by (a) smoothness of the concrete surface; (b) moisture content of the concrete; (c) type of coarse aggregate; (d) size, shape, and rigidity of specimen (e.g., a thick wall or beam); and (e) carbonation of the concrete surface.

b. Penetration resistance (probe).

(1) Description.

(a) The apparatus most often used for penetration resistance is the Windsor Probe, a special gun (Figure 2-36) that uses a 0.32 caliber blank with a precise quantity of powder to fire a high-strength steel probe into the concrete. A series of three measurements is made in each area with the spacer plate shown in Figure 2-37. The length of a probe extending from the surface of the concrete can be measured with a simple device, as shown in Figure 2-38.

(b) The manufacturer supplies a set of five calibration curves, each corresponding to a specific Moh's hardness for the coarse aggregate used in the concrete. With these curves, probe measurements are intended to be

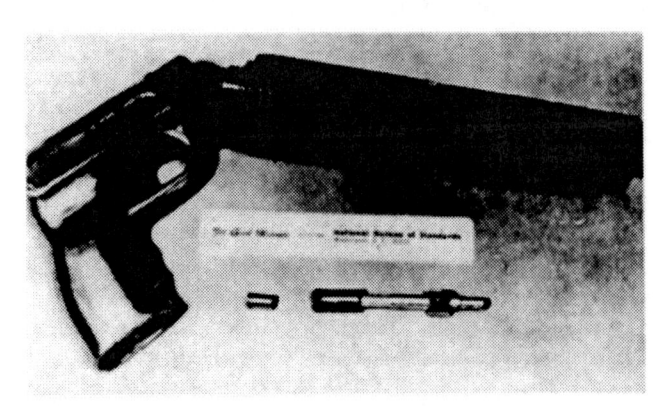

Figure 2-36. Windsor probe apparatus showing the gun, probe, and blank cartridge

Figure 2-37. Windsor probe in use

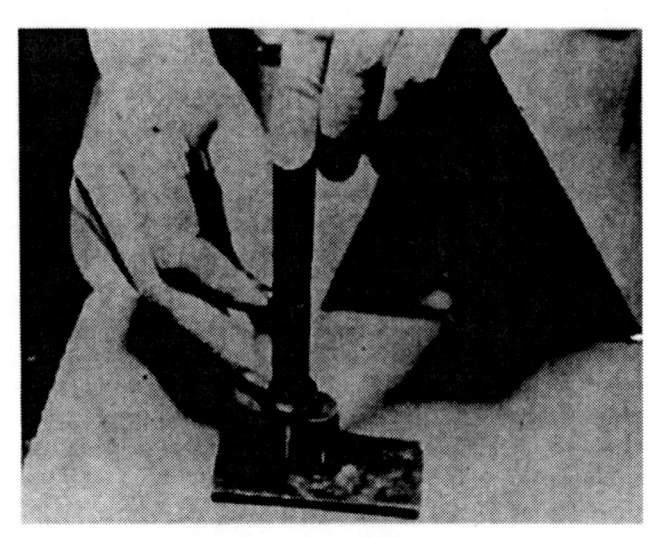

Figure 2-38. Device for measuring length of probe extending from surface of concrete

converted to compressive strength values. However, use of the manufacturer's calibration curves often results in grossly incorrect estimates of the compressive strength of concrete. Therefore, the penetration probe should be calibrated by the individual user and should be recalibrated whenever the type of aggregate or mixture is changed.

(2) Applications. Penetration resistance can be used for assessing the quality and uniformity of concrete because physical differences in concrete will affect its resistance to penetration. A probe will penetrate deeper as the density, subsurface hardness, and strength of the concrete decrease. Areas of poor concrete can be delineated by making a series of penetration tests at regularly spaced locations. The test method is covered in ASTM C 803 (CRD-C 59).

(3) Advantages. The probe equipment is simple, durable, requires little maintenance, and can be used by inspectors in the field with little training. The probe test is very useful in assessing the general quality and relative strength of concrete in different parts of a structure.

(4) Limitations. Care must be exercised whenever this device is used because a projectile is being fired; safety glasses should always be worn. The probe primarily measures surface and subsurface hardness; it does not yield precise measurements of the in situ strength of concrete. However, useful estimates of the compressive strength of concrete may be obtained if the probe is properly calibrated. The probe test does damage the concrete, leaving a hole of about 8 mm (0.32 in.) in diameter for the depth of the probe, and it may cause minor cracking and some surface spalling. Minor repairs of exposed surfaces may be necessary.

c. *Ultrasonic pulse-velocity method.*

(1) Description. The ultrasonic pulse-velocity method is probably the most widely used method for the nondestructive evaluation of in situ concrete. The method involves measurement of the time of travel of electronically pulsed compressional waves through a known

distance in concrete. From known TOA and distance traveled, the pulse velocity through the concrete can be calculated. Pulse-velocity measurements made through good-quality, continuous concrete will normally produce high velocities accompanied by good signal strengths. Poor-quality or deteriorated concrete will usually decrease velocity and signal strength. Concrete of otherwise good quality, but containing cracks, may produce high or low velocities, depending upon the nature and number of cracks but will almost always diminish signal strength.

(2) Applications. The ultrasonic pulse-velocity method has been used over the years to determine the general condition and quality of concrete, to assess the extent and severity of cracks in concrete, and to delineate areas of deteriorated or poor-quality concrete. The test method is described in ASTM C 597 (CRD-C 51).

(3) Advantages. The required equipment is portable (Figure 2-39) and has sufficient power to penetrate about 11 m (35 ft) of good continuous concrete, and the test can be performed quickly.

(4) Limitations. This method does not provide a precise estimate of concrete strength. Moisture variations and the presence of reinforcing steel can affect the results. Skilled personnel is required to analyze the results. The measurement requires access to opposite sides of the section being tested.

d. Acoustic mapping system.

(1) Description. This system makes possible, without dewatering of the structure, comprehensive evaluation of

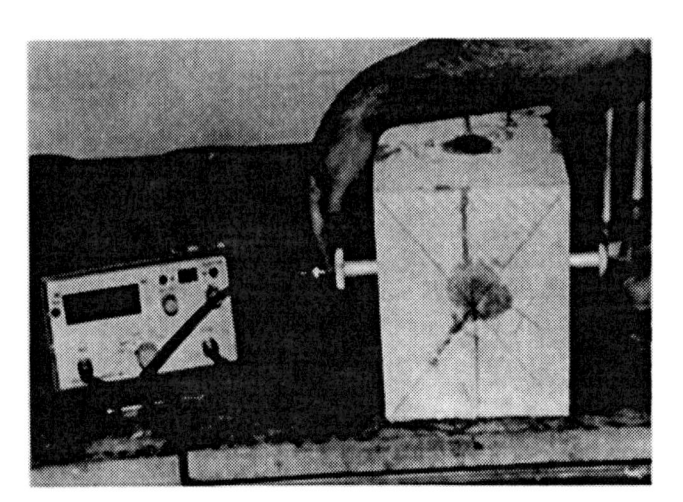

Figure 2-39. Ultrasonic pulse-velocity apparatus

top surface wear on such horizontal surfaces as aprons, sills, lock chamber floors, and stilling basins, where turbulent flows carrying rock and debris can cause abrasion-erosion damage. The system uses the sonar principle, i.e., transmitting acoustic waves and receiving reflections from underwater structures.

(2) Application. The system can be used to perform rapid, accurate surveys of submerged horizontal surfaces in water depths of 71.5 to 12 m (5 to 40 ft) with accuracies of ± 50 mm (2 in.) vertically and ± .3 m (1 ft) laterally. Variations of the system may be used for other underwater applications such as repairing and investigating large scour holes or silt buildup. The system has been successfully used in surveying the stilling basin floor of Folsom Dam, a U.S. Bureau of Reclamation project (SONEX 1984), and the stilling basin of Ice Harbor Dam in Walla Walla District (SONEX 1983).

(3) Advantages. This method avoids the expense and user inconvenience associated with dewatering and the dangers and inaccuracies inherent in diver-performed surveys.

(4) Limitations. Vertical and lateral accuracy will decrease at depths greater than 9 m (30 ft). There are some operational restrictions associated with water velocity and turbulence.

e. Ultrasonic pulse-echo (UPE).

(1) Description. A variation of the pulse-velocity technique is the pulse-echo method wherein a compressional wave pulse is transmitted from a surface and its echo received back at the same surface location. Reflection times from interfaces, cracks, or voids, together with the known velocity within the concrete, permit calculation of distances from the discontinuity to the transmitting and receiving points. The system has been demonstrated to be feasible but is still under development (Alexander and Thornton 1988). An impact pulse-echo system for measurements on concrete piles is described by Alexander (1980).

(2) Applications. The system operates well for flatwork for dimensions less than 0.3 m (1 ft) in thickness. The system can detect foreign objects such as steel and plastic pipe. It can measure unknown thicknesses and presence of delaminations up to 0.3 m (1 ft) in thickness. Recently neural network algorithms were trained on some calibrated specimens to recognize the condition of concrete that has uniform microcracking.

(3) Advantages. The system has excellent resolution as it operates around a center frequently of 200 kHz. The wavelength is roughly 25 mm (1 in.) long in good-quality concrete, which provides better spatial resolution than radar. It can operate underwater or in the dry. The speed of sound in concrete does not vary by more than 5 percent from moist to dry concrete.

(4) Limitations. Presently the system exists as a laboratory prototype. The equipment presently is multi-component and not very portable. Also, most measurement data need digital signal algorithms applied to the data to bring signals out of the noise, and this task requires the expertise of someone skilled in that discipline. The system presently does not have an onboard computer, and the data cannot be processed onsite in realtime. The system is not yet available commercially and is not a CRD or ASTM measurement standard. Plans are underway to commercialize the system and remedy the above-mentioned limitations.

f. Radar.

(1) Description. This is a reflection technique that is based on the principle of electromagnetic wave propagation. Similar to UPE in operation, the TOA of the wave is measured from the time the pulse is introduced into the concrete at the surface of the structure, travels to the discontinuity or interface, and is reflected back to original surface. Whereas the mechanical wave travels at the speed of sound for the UPE technique, the electromagnetic wave travels at the speed of light for radar.

(2) Applications. A radar unit operating at the frequency of 1 gHz has a wavelength about 150 mm (6 in.) in concrete. Presently systems can penetrate to a depth of about 0.5 m (1.5 ft) at this frequency. A void 150 mm (6 in.) deep in concrete must have a diameter of 50 to 75 mm (2 to 3 in.) to be detectable. At a depth of 0.3 m (1 ft), the void must be 75 to 100 mm (3 to 4 in.) in diameter to be detectable. Lower frequency systems can penetrate deeper than this, but the resolution is even poorer. Radar is especially sensitive for detecting steel reinforcement, but steel can also interfere with the measurements if one is looking for deterioration in the concrete. Radar is sensitive to moisture and may be useful for finding deteriorated areas, which tend to hold more water than sound concrete.

(3) Advantages. Radar is a noncontact method and data acquisition is very fast. Resolution and penetration are limited at the present time. Systems are available commercially.

(4) Limitations. Radar is steel in the process of development for use on concrete (Ahmad and Haskins 1993), and a measurement standard does not exist at this time. A radar unit may cost between \$50K and \$100K and requires someone highly trained to operate the equipment and interpret the data. Commercial systems being used for concrete are primarily designed to operate in the earth for geophysical applications. Better results can be obtained by applying signal processing techniques. The velocity of the pulse is dependent on the dielectric constant of the concrete and varies by almost 100 percent between dry concrete and moist concrete.

2-7. Stability Analysis

A stability analysis is often performed as part of an overall evaluation of the condition of a concrete structure. Guidelines for performing a stability analysis for existing structures are beyond the scope of this manual, but may be found in other CE publications. Information on requirements for stability analyses may be obtained from CECW-E.

2-8. Deformation Monitoring

A tool now available for a comprehensive evaluation of larger structures is the Continuous Deformation Monitoring System (CDMS) developed in Repair, Evaluation, Maintenance, and Rehabilitation (REMR) Research Program. The CDMS uses the Navigation Satellite Timing and Ranging (NAVSTAR) Global Positioning Systems (GPS) to monitor the position of survey monuments installed on a structure. The system was demonstrated in a field test at Dworshak Dam (Lanigan 1992).

2-9. Concrete Service Life

a. Freeze-thaw deterioration. A procedure has been developed to predict the service life of nonair-entrained concrete subject to damage from freezing and thawing. The procedure addresses with a probabilistic method (Bryant and Mlakar 1991) both the known and uncertain qualities of the relevant material properties, environmental factors, and model of degradation resulting from freezing and thawing. Two important characteristics of this procedure are (1) it rationally addresses the uncertainties inherent in degradation of mass concrete caused by freezing and thawing, and (2) it is mathematically straightforward for implementation by CE offices.

(1) Current procedures for thermal modeling and analysis appear quite adequate for predicting temperatures in a concrete structure. Although 2-D analyses are better

for determining complex thermal response, in many cases a series of much simpler one-dimensional (1-D) analyses provide a very good estimation of temperatures. The external temperature inputs to a thermal analysis, i.e., water-air temperatures, were well represented by sinusoidal curves.

(2) The general understanding and analytical models for predicting moisture migration and degree of saturation are not as well developed as those for the thermal problem. A seepage model for predicting the degree of saturation appears to provide adequate answers for the prediction of service life; however, further study is appropriate to substantiate this indication.

(3) The procedure was demonstrated by hindcast application to the middle wall and landwall at Dashields Lock which exhibited an appreciable degree of measurable damage caused by freezing and thawing. Required data for application of the procedure, e.g., temperature and concrete properties, were available for these features, which were representative of typical CE projects.

(4) Damage predicted by the procedure was in agreement with observed damage resulting from freezing and thawing at each site. The general trends of location and spatial variation of damage were very similar to observations and measurements at the two sites. More encouragingly, the actual magnitudes of damage predicted by the procedure compared favorably with the previous measurements. This result provides the strongest indication that the procedure is rational and would enhance the ability of the CE to predict service life at its many other concrete structures.

b. Other deterioration mechanisms. A complete and comprehensive report by Clifton (1991) examines the basis for predicting the remaining service lives of concrete materials of nuclear power facilities. The study consisted of two major activities: the evaluation of models which can be used in predicting the remaining service life of concrete exposed to the major environmental stressors and aging factors; and, the evaluation of accelerated aging techniques and tests which can provide data for service life models or which themselves can be used to predict the remaining service life of concrete. Methods for service life prediction which are discussed in this report include: (1) estimates based on experience; (2) deductions from performance of similar materials; (3) accelerated testing; (4) applications of reliability and stochastic concepts; and (5) mathematical modeling based on the chemistry and physics of the degradation processes. Models for corrosion, sulfate attack, frost attack, and leaching were identified and analyzed. While no model was identified for distress caused by alkali-aggregate reactions, an approach for modeling the process was outlined.

2-10. Reliability Analysis

A reliability analysis may be required for major rehabilitation projects. Guidelines for performing a reliability analysis are beyond the scope of this manual. Information on requirements for reliability analyses may be obtained from CECW-E.

Chapter 3
Causes of Distress and Deterioration of Concrete

3-1. Introduction

a. General. Once the evaluation phase has been completed for a structure, the next step is to establish the cause or causes for the damage that has been detected. Since many of the symptoms may be caused by more than one mechanism acting upon the concrete, it is necessary to have an understanding of the basic underlying causes of damage and deterioration. This chapter presents information on the common causes of problems in concrete. These causes are shown in Table 3-1. Items shown in the table are discussed in the subsequent sections of this chapter with the following given for each: (1) brief discussion of the basic mechanism; (2) description of the most typical symptoms, both those that would be observed during a visual examination and those that would be seen during a laboratory evaluation; and (3) recommendations for preventing further damage to new or replacement concrete. The last section of the chapter presents a logical method for relating the symptoms or observations to the various causes.

b. Approach to evaluation. Deterioration of concrete is an extremely complex subject. It would be simplistic to suggest that it will be possible to identify a specific, single cause of deterioration for every symptom detected during an evaluation of a structure. In most cases, the damage detected will be the result of more than one mechanism. For example, corrosion of reinforcing steel may open cracks that allow moisture greater access to the interior of the concrete. This moisture could lead to additional damage by freezing and thawing. In spite of the complexity of several causes working simultaneously, given a basic understanding of the various damage-causing mechanisms, it should be possible, in most cases, to determine the primary cause or causes of the damage seen on a particular structure and to make intelligent choices concerning selection of repair materials and methods.

3-2. Causes of Distress and Deterioration

a. Accidental loadings.

(1) Mechanism. Accidental loadings may be characterized as short-duration, one-time events such as the impact of a barge against a lock wall or an earthquake.

Table 3-1
Causes of Distress and Deterioration of Concrete

Accidental Loadings

Chemical Reactions

  Acid attack

  Aggressive-water attack

  Alkali-carbonate rock reaction

  Alkali-silica reaction

  Miscellaneous chemical attack

  Sulfate attack

Construction Errors

Corrosion of Embedded Metals

Design Errors

  Inadequate structural design

  Poor design details

Erosion

  Abrasion

  Cavitation

Freezing and Thawing

Settlement and Movement

Shrinkage

  Plastic

  Drying

Temperature Changes

  Internally generated

  Externally generated

  Fire

Weathering

These loadings can generate stresses higher than the strength of the concrete, resulting in localized or general failure. Determination of whether accidental loading caused damage to the concrete will require knowledge of the events preceding discovery of the damage. Usually, damage caused by accidental loading will be easy to diagnose.

(2) Symptoms. Visual examination will usually show spalling or cracking of concrete which has been subjected to accidental loadings. Laboratory analysis is generally not necessary.

(3) Prevention. Accidental loadings by their very nature cannot be prevented. Minimizing the effects of some occurrences by following proper design procedures (an example is the design for earthquakes) or by proper

attention to detailing (wall armor in areas of likely impact) will reduce the impacts of accidental loadings.

b. Chemical reactions. This category includes several specific causes of deterioration that exhibit a wide variety of symptoms. In general, deleterious chemical reactions may be classified as those that occur as the result of external chemicals attacking the concrete (acid attack, aggressive water attack, miscellaneous chemical attack, and sulfate attack) or those that occur as a result of internal chemical reactions between the constituents of the concrete (alkali-silica and alkali-carbonate rock reactions). Each of these chemical reactions is described below.

(1) Acid attack.

(a) Mechanism. Portland-cement concrete is a highly alkaline material and is not very resistant to attack by acids. The deterioration of concrete by acids is primarily the result of a reaction between the acid and the products of the hydration of cement. Calcium silicate hydrate may be attacked if highly concentrated acid exists in the environment of the concrete structures. In most cases, the chemical reaction results in the formation of water-soluble calcium compounds that are then leached away. In the case of sulfuric acid attack, additional or accelerated deterioration results because the calcium sulfate formed may affect the concrete by the sulfate attack mechanism (Section 3-2*b*(6)). If the acid is able to reach the reinforcing steel through cracks or pores in the concrete, corrosion of the reinforcing steel will result and will cause further deterioration of the concrete (ACI 201.2R).

(b) Symptoms. Visual examination will show disintegration of the concrete evidenced by loss of cement paste and aggregate from the matrix (Figure 2-13). If reinforcing steel has been reached by the acid, rust staining, cracking, and spalling may be present. If the nature of the solution in which the deteriorating concrete is located is unknown, laboratory analysis can be used to identify the specific acid involved.

(c) Prevention. A dense concrete with a low water-cement ratio (w/c) may provide an acceptable degree of protection against a mild acid attack. Portland-cement concrete, because of its composition, is unable to withstand attack by highly acidic solutions for long periods of time. Under such conditions, an appropriate surface coating or treatment may be necessary. ACI Committee 515 has extensive recommendations for such coatings (ACI 515.1R).

(2) Aggressive-water attack.

(a) Mechanism. Some waters have been reported to have extremely low concentrations of dissolved minerals. These soft or aggressive waters will leach calcium from cement paste or aggregates. This phenomenon has been infrequently reported in the United States. From the few cases that have been reported, there are indications that this attack takes place very slowly. For an aggressive-water attack to have a serious effect on hydraulic structures, the attack must occur in flowing water. This keeps a constant supply of aggressive water in contact with the concrete and washes away aggregate particles that become loosened as a result of leaching of the paste (Holland, Husbands, Buck, and Wong 1980).

(b) Symptoms. Visual examination will show concrete surfaces that are very rough in areas where the paste has been leached (Figure 2-12). Sand grains may be present on the surface of the concrete, making it resemble a coarse sandpaper. If the aggregate is susceptible to leaching, holes where the coarse aggregate has been dissolved will be evident. Water samples from structures where aggressive-water attack is suspected may be analyzed to calculate the Langlier Index, which is a measure of the aggressiveness of the water (Langlier 1936).

(c) Prevention. The aggressive nature of water at the site of a structure can be determined before construction or during a major rehabilitation. Additionally, the water-quality evaluation at many structures can be expanded to monitor the aggressiveness of water at the structure. If there are indications that the water is aggressive or is becoming aggressive, areas susceptible to high flows may be coated with a nonportland-cement-based coating.

(3) Alkali-carbonate rock reaction.

(a) Mechanism. Certain carbonate rock aggregates have been reactive in concrete. The results of these reactions have been characterized as ranging from beneficial to destructive. The destructive category is apparently limited to reactions with impure dolomitic aggregates and are a result of either dedolomitization or rim-silicification reactions. The mechanism of alkali-carbonate rock reaction is covered in detail in EM 1110-2-2000.

(b) Symptoms. Visual examination of those reactions that are serious enough to disrupt the concrete in a structure will generally show map or pattern cracking and a general appearance which indicates that the concrete is swelling (Figure 2-7). A distinguishing feature which differentiates alkali-carbonate rock reaction from alkali-silica reaction is the lack of silica gel exudations at cracks (ACI 201.2R). Petrographic examination in accordance with ASTM C 295 (CRD-C 127) may be used to confirm the presence of alkali-carbonate rock reaction.

(c) Prevention. In general, the best prevention is to avoid using aggregates that are or suspected of being reactive. Appendix E of EM 1110-2-2000 prescribes procedures for testing rocks for reactivity and for minimizing effects when reactive aggregates must be used.

(4) Alkali-silica reaction.

(a) Mechanism. Some aggregates containing silica that is soluble in highly alkaline solutions may react to form a solid nonexpansive calcium-alkali-silica complex or an alkali-silica complex which can imbibe considerable amounts of water and then expand, disrupting the concrete. Additional details may be found in EM 1110-2-2000.

(b) Symptoms. Visual examination of those concrete structures that are affected will generally show map or pattern cracking and a general appearance that indicates that the concrete is swelling (Figure 2-6). Petrographic examination may be used to confirm the presence of alkali-silica reaction.

(c) Prevention. In general, the best prevention is to avoid using aggregates that are known or suspected to be reactive or to use a cement containing less than 0.60 percent alkalies (percent Na_2O + (0.658) percent K_2O). Appendix D of EM 1110-2-2000 prescribes procedures for testing aggregates for reactivity and for minimizing the effects when reactive aggregates must be used.

(5) Miscellaneous chemical attack.

(a) Mechanism. Concrete will resist chemical attack to varying degrees, depending upon the exact nature of the chemical. ACI 515.1R includes an extensive listing of the resistance of concrete to various chemicals. To produce significant attack on concrete, most chemicals must be in solution that is above some minimum concentration. Concrete is seldom attacked by solid dry chemicals. Also, for maximum effect, the chemical solution needs to be circulated in contact with the concrete. Concrete subjected to aggressive solutions under positive differential pressure is particularly vulnerable. The pressure gradients

tend to force the aggressive solutions into the matrix. If the low-pressure face of the concrete is exposed to evaporation, a concentration of salts tends to accumulate at that face, resulting in increased attack. In addition to the specific nature of the chemical involved, the degree to which concrete resists attack depends upon the temperature of the aggressive solution, the w/c of the concrete, the type of cement used (in some circumstances), the degree of consolidation of the concrete, the permeability of the concrete, the degree of wetting and drying of the chemical on the concrete, and the extent of chemically induced corrosion of the reinforcing steel (ACI 201.1R).

(b) Symptoms. Visual examination of concrete which has been subjected to chemical attack will usually show surface disintegration and spalling and the opening of joints and cracks. There may also be swelling and general disruption of the concrete mass. Coarse aggregate particles are generally more inert than the cement paste matrix; therefore, aggregate particles may be seen as protruding from the matrix. Laboratory analysis may be required to identify the unknown chemicals which are causing the damage.

(c) Prevention. Typically, dense concretes with low w/c (maximum w/c = 0.40) provide the greatest resistance. The best known method of providing long-term resistance is to provide a suitable coating as outlined in ACI 515.1R.

(6) Sulfate attack.

(a) Mechanism. Naturally occurring sulfates of sodium, potassium, calcium, or magnesium are sometimes found in soil or in solution in ground water adjacent to concrete structures. The sulfate ions in solution will attack the concrete. There are apparently two chemical reactions involved in sulfate attack on concrete. First, the sulfate reacts with free calcium hydroxide which is liberated during the hydration of the cement to form calcium sulfate (gypsum). Next, the gypsum combines with hydrated calcium aluminate to form calcium sulfoaluminate (ettringite). Both of these reactions result in an increase in volume. The second reaction is mainly responsible for most of the disruption caused by volume increase of the concrete (ACI 201.2R). In addition to the two chemical reactions, there may also be a purely physical phenomenon in which the growth of crystals of sulfate salts disrupts the concrete.

(b) Symptoms. Visual examination will show map and pattern cracking as well as a general disintegration of the concrete (Figure 2-14). Laboratory analysis can verify the occurrence of the reactions described.

(c) Prevention. Protection against sulfate attack can generally be obtained by the following: Use of a dense, high-quality concrete with a low water-cement ratio; Use of either a Type V or a Type II cement, depending upon the anticipated severity of the exposure (EM 1110-2-2000); Use of a suitable pozzolan (some pozzolans, added as part of a blended cement or separately, have improved resistance, while others have hastened deterioration). If use of a pozzolan is anticipated, laboratory testing to verify the degree of improvement to be expected is recommended.

c. Construction errors. Failure to follow specified procedures and good practice or outright carelessness may lead to a number of conditions that may be grouped together as construction errors. Typically, most of these errors do not lead directly to failure or deterioration of concrete. Instead, they enhance the adverse impacts of other mechanisms identified in this chapter. Each error will be briefly described below along with preventative methods. In general, the best preventive measure is a thorough knowledge of what these construction errors are plus an aggressive inspection program. It should be noted that errors of the type described in this section are equally as likely to occur during repair or rehabilitation projects as they are likely to occur during new construction.

(1) Adding water to concrete. Water is usually added to concrete in one or both of the following circumstances: First, water is added to the concrete in a delivery truck to increase slump and decrease emplacement effort. This practice will generally lead to concrete with lowered strength and reduced durability. As the w/c of the concrete increases, the strength and durability will decrease. In the second case, water is commonly added during finishing of flatwork. This practice leads to scaling, crazing, and dusting of the concrete in service.

(2) Improper alignment of formwork. Improper alignment of the formwork will lead to discontinuities on the surface of the concrete. While these discontinuities are unsightly in all circumstances, their occurrence may be more critical in areas that are subjected to high-velocity flow of water, where cavitation-erosion may be induced, or in lock chambers where the "rubbing" surfaces must be straight.

(3) Improper consolidation. Improper consolidation of concrete may result in a variety of defects, the most common being bugholes, honeycombing, and cold joints.

"Bugholes" are formed when small pockets of air or water are trapped against the forms. A change in the mixture to make it less "sticky" or the use of small vibrators worked near the form has been used to help eliminate bugholes. Honeycombing can be reduced by inserting the vibrator more frequently, inserting the vibrator as close as possible to the form face without touching the form, and slower withdrawal of the vibrator. Obviously, any or all of these defects make it much easier for any damage-causing mechanism to initiate deterioration of the concrete. Frequently, a fear of "overconsolidation" is used to justify a lack of effort in consolidating concrete. Overconsolidation is usually defined as a situation in which the consolidation effort causes all of the coarse aggregate to settle to the bottom while the paste rises to the surface. If this situation occurs, it is reasonable to conclude that there is a problem of a poorly proportioned concrete rather than too much consolidation.

(4) Improper curing. Curing is probably the most abused aspect of the concrete construction process. Unless concrete is given adequate time to cure at a proper humidity and temperature, it will not develop the characteristics that are expected and that are necessary to provide durability. Symptoms of improperly cured concrete can include various types of cracking and surface disintegration. In extreme cases where poor curing leads to failure to achieve anticipated concrete strengths, structural cracking may occur.

(5) Improper location of reinforcing steel. This section refers to reinforcing steel that is improperly located or is not adequately secured in the proper location. Either of these faults may lead to two general types of problems. First, the steel may not function structurally as intended, resulting in structural cracking or failure. A particularly prevalent example is the placement of welded wire mesh in floor slabs. In many cases, the mesh ends up on the bottom of the slab which will subsequently crack because the steel is not in the proper location. The second type of problem stemming from improperly located or tied reinforcing steel is one of durability. The tendency seems to be for the steel to end up near the surface of the concrete. As the concrete cover over the steel is reduced, it is much easier for corrosion to begin.

(6) Movement of formwork. Movement of formwork during the period while the concrete is going from a fluid to a rigid material may induce cracking and separation within the concrete. A crack open to the surface will allow access of water to the interior of the concrete. An internal void may give rise to freezing or corrosion problems if the void becomes saturated.

(7) Premature removal of shores or reshores. If shores or reshores are removed too soon, the concrete affected may become overstressed and cracked. In extreme cases there may be major failures.

(8) Settling of the concrete. During the period between placing and initial setting of the concrete, the heavier components of the concrete will settle under the influence of gravity. This situation may be aggravated by the use of highly fluid concretes. If any restraint tends to prevent this settling, cracking or separations may result. These cracks or separations may also develop problems of corrosion or freezing if saturated.

(9) Settling of the subgrade. If there is any settling of the subgrade during the period after the concrete begins to become rigid but before it gains enough strength to support its own weight, cracking may also occur.

(10) Vibration of freshly placed concrete. Most construction sites are subjected to vibration from various sources, such as blasting, pile driving, and from the operation of construction equipment. Freshly placed concrete is vulnerable to weakening of its properties if subjected to forces which disrupt the concrete matrix during setting. The vibration limits for concrete, expressed in terms of peak particle velocity and given in Table 3-2, were established as a result of laboratory and field test programs.

(11) Improper finishing of flat work. The most common improper finishing procedures which are detrimental to the durability of flat work are discussed below.

(a) Adding water to the surface. This procedure was discussed in paragraph 3-2c(1) above. Evidence that water is being added to the surface is the presence of a large paint brush, along with other finishing tools. The brush is dipped in water and water is "slung" onto the surface being finished.

(b) Timing of finishing. Final finishing operations must be done after the concrete has taken its initial set and bleeding has stopped. The waiting period depends on the amounts of water, cement, and admixtures in the mixture but primarily on the temperature of the concrete surface. On a partially shaded slab, the part in the sun will usually be ready to finish before the part in the shade.

(c) Adding cement to the surface. This practice is often done to dry up bleed water to allow finishing to proceed and will result in a thin cement-rich coating which will craze or flake off easily.

(d) Use of tamper. A tamper or "jitterbug" is unnecessarily used on many jobs. This tool forces the coarse aggregate away from the surface and can make finishing easier. This practice, however, creates a cement-rich mortar surface layer which can scale or craze. A jitterbug should not be allowed with a well designed mixture. If a harsh mixture must be finished, the judicious use of a jitterbug could be useful.

(e) Jointing. The most frequent cause of cracking in flatwork is the incorrect spacing and location of joints. Joint spacing is discussed in ACI 330R.

d. Corrosion of embedded metals.

(1) Mechanisms. Steel reinforcement is deliberately and almost invariably placed within a few inches of a concrete surface. Under most circumstances, portland-cement concrete provides good protection to the embedded reinforcing steel. This protection is generally attributed to the high alkalinity of the concrete adjacent to the steel and to the relatively high electrical resistance of the concrete. Still, corrosion of the reinforcing steel is among the most frequent causes of damage to concrete.

Table 3-2
Vibration Limits for Freshly Placed Concrete (Hulshizer and Desci 1984)

Age of Concrete at Time of Vibration (hr)	Peak Particle Velocity of Ground Vibrations
Up to 3	102 mm/sec (4.0 in./sec)
3 to 11	38 mm/sec (1.5 in./sec)
11 to 24	51 mm/sec (2.0 in./sec)
24 to 48	102 mm/sec (4.0 in./sec)
Over 48	178 mm/sec (7.0 in./sec)

(a) High alkalinity and electrical resistivity of the concrete. The high alkalinity of the concrete pore solution can be reduced over a long period of time by carbonation. The electrical resistivity can be decreased by the presence of chemicals in the concrete. The chemical most commonly applied to concrete is chloride salts in the form of deicers. As the chloride ions penetrate the concrete, the capability of the concrete to carry an electrical current is increased significantly. If there are differences within the concrete such as moisture content, chloride content, oxygen content, or if dissimilar metals are in contact, electrical potential differences will occur and a corrosion cell may be established. The anodes will experience corrosion while the cathodes will be undamaged. On an individual reinforcing bar there may be many anodes and cathodes, some adjacent, and some widely spaced.

(b) Corrosion-enhanced reduction in load-carrying capacity of concrete. As the corrosion progresses, two things occur: First, the cross-sectional area of the reinforcement is reduced, which in turn reduces the load-carrying capacity of the steel. Second, the products of the corrosion, iron oxide (rust), expand since they occupy about eight times the volume of the original material. This increase in volume leads to cracking and ultimately spalling of the concrete. For mild steel reinforcing, the damage to the concrete will become evident long before the capacity of the steel is reduced enough to affect its load-carrying capacity. However, for prestressing steel, slight reductions in section can lead to catastrophic failure.

(c) Other mechanisms for corrosion of embedded metals. In addition to the development of an electrolytic cell, corrosion may be developed under several other situations. The first of these is corrosion produced by the presence of a stray electrical current. In this case, the current necessary for the corrosion reaction is provided from an outside source. A second additional source of corrosion is that produced by chemicals that may be able to act directly on the reinforcing steel. Since this section has dealt only with the corrosion of steel embedded in concrete, for information on the behavior of other metals in concrete, see ACI 201.2R and ACI 222R.

(2) Symptoms. Visual examination will typically reveal rust staining of the concrete. This staining will be followed by cracking. Cracks produced by corrosion generally run in straight, parallel lines at uniform intervals corresponding to the spacing of the reinforcement. As deterioration continues, spalling of the concrete over the reinforcing steel will occur with the reinforcing bars becoming visible (Figure 2-27). One area where laboratory analysis may be beneficial is the determination of the chloride contents in the concrete. This procedure may be used to determine the amount of concrete to be removed during a rehabilitation project.

(3) Prevention. ACI 201.2R describes the considerations for protecting reinforcing steel in concrete: use of concrete with low permeability; use of properly proportioned concrete having a low w/c; use of as low a concrete slump as practical; use of good workmanship in placing the concrete; curing the concrete properly; providing adequate concrete cover over the reinforcing steel; providing good drainage to prevent water from standing on the concrete; limiting chlorides in the concrete mixture; and paying careful attention to protruding items such as bolts or other anchors.

e. Design errors. Design errors may be divided into two general types: those resulting from inadequate structural design and those resulting from lack of attention to relatively minor design details. Each of the two types of design errors is discussed below.

(1) Inadequate structural design.

(a) Mechanism. The failure mechanism is simple-- the concrete is exposed to greater stress than it is capable of carrying or it sustains greater strain than its strain capacity.

(b) Symptoms. Visual examinations of failures resulting from inadequate structural design will usually show one of two symptoms. First, errors in design resulting in excessively high compressive stresses will result in spalling. Similarly, high torsion or shear stresses may also result in spalling or cracking. Second, high tensile stresses will result in cracking. To identify inadequate design as a cause of damage, the locations of the damage should be compared to the types of stresses that should be present in the concrete. For example, if spalls are present on the underside of a simple-supported beam, high compressive stresses are not present and inadequate design may be eliminated as a cause. However, if the type and location of the damage and the probable stress are in agreement, a detailed stress analysis will be required to determine whether inadequate design is the cause. Laboratory analysis is generally not applicable in the case of suspected inadequate design. However, for rehabilitation projects, thorough petrographic analysis and strength testing of concrete from elements to be reused will be necessary.

(c) Prevention. Inadequate design is best prevented by thorough and careful review of all design calculations. Any rehabilitation method that makes use of existing concrete structural members must be carefully reviewed.

(2) Poor design details. While a structure may be adequately designed to meet loadings and other overall requirements, poor detailing may result in localized concentrations of high stresses in otherwise satisfactory concrete. These high stresses may result in cracking that allows water or chemicals access to the concrete. In other cases, poor design detailing may simply allow water to pond on a structure, resulting in saturated concrete. In general, poor detailing does not lead directly to concrete failure; rather, it contributes to the action of one of the other causes of concrete deterioration described in this chapter. Several specific types of poor detailing and their possible effects on a structure are described in the following paragraphs. In general, all of these problems can be prevented by a thorough and careful review of plans and specifications for the project. In the case of existing structures, problems resulting from poor detailing should be handled by correcting the detailing and not by simply responding to the symptoms.

(a) Abrupt changes in section. Abrupt changes in section may cause stress concentrations that may result in cracking. Typical examples would include the use of relatively thin sections such as bridge decks rigidly tied into massive abutments or patches and replacement concrete that are not uniform in plan dimensions.

(b) Insufficient reinforcement at reentrant corners and openings. Reentrant corners and openings also tend to cause stress concentrations that may cause cracking. In this case, the best prevention is to provide additional reinforcement in areas where stress concentrations are expected to occur.

(c) Inadequate provision for deflection. Deflections in excess of those anticipated may result in loading of members or sections beyond the capacities for which they were designed. Typically, these loadings will be induced in walls or partitions, resulting in cracking.

(d) Inadequate provision for drainage. Poor attention to the details of draining a structure may result in the ponding of water. This ponding may result in leakage or saturation of concrete. Leakage may result in damage to the interior of the structure or in staining and encrustations on the structure. Saturation may result in severely damaged concrete if the structure is in an area that is subjected to freezing and thawing.

(e) Insufficient travel in expansion joints. Inadequately designed expansion joints may result in spalling of concrete adjacent to the joints. The full range of possible temperature differentials that a concrete may be expected to experience should be taken into account in the specification for expansion joints. There is no single expansion joint that will work for all cases of temperature differential.

(f) Incompatibility of materials. The use of materials with different properties (modulus of elasticity or coefficient of thermal expansion) adjacent to one another may result in cracking or spalling as the structure is loaded or as it is subjected to daily or annual temperature variations.

(g) Neglect of creep effect. Neglect of creep may have similar effects as noted earlier for inadequate provision for deflections (paragraph 3-2e(2)(c)). Additionally, neglect of creep in prestressed concrete members may lead to excessive prestress loss that in turn results in cracking as loads are applied.

(h) Rigid joints between precast units. Designs utilizing precast elements must provide for movement between adjacent precast elements or between the precast elements and the supporting frame. Failure to provide for this movement can result in cracking or spalling.

(i) Unanticipated shear stresses in piers, columns, or abutments. If, through lack of maintenance, expansion bearing assembles are allowed to become frozen, horizontal loading may be transferred to the concrete elements supporting the bearings. The result will be cracking in the concrete, usually compounded by other problems which will be caused by the entry of water into the concrete.

(j) Inadequate joint spacing in slabs. This is one of the most frequent causes of cracking of slabs-on-grade. Guidance on joint spacing and depth of contraction joints may be found in ACI 332R.

f. Abrasion. Abrasion damage caused by water-borne debris and the techniques used to repair the damage on several Corps' structures are described by McDonald (1980). Also, causes of abrasion-erosion damage and procedures for repair and prevention of damage are described in ACI 210R.

(1) Mechanism. Abrasion-erosion damage is caused by the action of debris rolling and grinding against a concrete surface. In hydraulic structures, the areas most

likely to be damaged are spillway aprons, stilling basin slabs, and lock culverts and laterals. The sources of the debris include construction trash left in a structure, riprap brought back into a basin by eddy currents because of poor hydraulic design or asymmetrical discharge, and riprap or other debris thrown into a basin by the public. Also barges and towboats impacting or scraping on lock wells and guide wells can cause abrasions erosion damage.

(2) Symptoms. Concrete surfaces abraded by water-borne debris are generally smooth (Figure 2-20) and may contain localized depressions. Most of the debris remaining in the structure will be spherical and smooth. Mechanical abrasion is usually characterized by long shallow grooves in the concrete surface and spalling along monolith joints. Armor plates is often torn away or bent.

(3) Prevention. The following measures should be followed to prevent or minimize abrasion-erosion damage to concrete hydraulic structures (Liu 1980 and McDonald 1980).

(a) Design. It appears that given appropriate flow conditions in the presence of debris, all of the construction materials currently being used in hydraulic structures are to some degree susceptible to erosion. While improvements in materials should reduce the rate of concrete damage caused by erosion, this improvement alone will not solve the problem. Until the adverse hydraulic conditions that can cause abrasion-erosion damage are minimized or eliminated, it will be extremely difficult for any of the construction materials currently being used to avoid damage by erosion. Prior to construction or repair of major structures, hydraulic model studies of the structure may be required to identify potential causes of erosion damage and to evaluate the effectiveness of various modifications in eliminating those undesirable hydraulic conditions. Many older structures have spillways designed with a vertical end-sill. This design is usually efficient in trapping the erosion-causing debris within the spillway. In some structures, a 45-deg fillet installed on the upstream side of the end sill has resulted in a self-cleaning stilling basin. Recessing monolith joints in lock walls and guide walls will minimize stilling basin spalling caused by barge impact and abrasion (See paragraph 8-1e(2)(e)).

(b) Operation. In existing structures, balanced flows should be maintained into basins by using all gates to avoid discharge conditions where eddy action is prevalent. Substantial discharges that can provide a good hydraulic jump without creating eddy action should be released periodically in an attempt to flush debris from the stilling basin. Guidance as to discharge and tailwater relations required for flushing should be developed through model and prototype tests. Periodic inspections should be required to determine the presence of debris in the stilling basin and the extent of erosion. If the debris cannot be removed by flushing operations, the basin should be cleaned by other means.

(c) Materials. It is imperative that materials be tested and evaluated, in accordance with ASTM C 1138 (CRD-C 63), prior to use in the repair of abrasion-erosion damaged hydraulic structures. Abrasion-resistant concrete should include the maximum amount of the hardest coarse aggregate that is available and the lowest practical w/c. In some cases where hard aggregate was not available, high-range water-reducing admixtures (HRWRA) and condensed silica fume have been used to develop high compressive strength concrete 97 MPa (14,000 psi) to overcome problems of unsatisfactory aggregate (Holland 1983). Apparently, at these high compressive strengths the hardened cement paste assumes a greater role in resisting abrasion-erosion damage, and as such, the aggregate quality becomes correspondingly less important. The abrasion-erosion resistance of vacuum-treated concrete, polymer concrete, polymer-impregnated concrete, and polymer portland-cement concrete is significantly superior to that of comparable conventional concrete that can also be attributed to a stronger cement matrix. The increased costs associated with materials, production, and placing of these and any other special concretes in comparison with conventional concrete should be considered during the evaluation process. While the addition of steel fibers would be expected to increase the impact resistance of concrete, fiber-reinforced concrete is consistently less resistant to abrasion-erosion than conventional concrete. Therefore, fiber-reinforced concrete should not be used for repair of stilling basins or other hydraulic structures where abrasion-erosion is of major concern. Several types of surface coatings have exhibited good abrasion-erosion resistance during laboratory tests. These include polyurethanes, epoxy-resin mortar, furan-resin mortar, acrylic mortar, and iron aggregate toppings. However, some difficulties have been reported in field applications of surface coatings, primarily the result of improper surface preparation and thermal incompatibility between coatings and concrete.

g. Cavitation. Cavitation-erosion is the result of relatively complex flow characteristics of water over concrete surfaces (ACI 210R).

(1) Mechanism. There is little evidence to show that water flowing over concrete surfaces at velocities less

than 12.2 m/sec (40 ft/sec) causes any cavitation damage to the concrete. However, when the flow is fast enough (greater than 12.2 m/sec) and where there is surface irregularity in the concrete, cavitation damage may occur. Whenever there is surface irregularity, the flowing water will separate from the concrete surface. In the area of separation from the concrete, vapor bubbles will develop because of the lowered vapor pressure in the region. As these bubbles are carried downstream, they will soon reach areas of normal pressure. These bubbles will collapse with an almost instantaneous reduction in volume. This collapse, or implosion, creates a shock wave which, upon reaching a concrete surface, induces very high stresses over a small area. The repeated collapse of vapor bubbles on or near the concrete surface will cause pitting. Concrete spillways and outlet works of many high dams have been severely damaged by cavitation.

(2) Symptoms. Concrete that has been damaged will be severely pitted and extremely rough (Figure 2-21). As the damage progresses, the roughness of the damaged area may induce additional cavitation.

(3) Prevention.

(a) Hydraulic design. Even the strongest materials cannot withstand the forces of cavitation indefinitely. Therefore, proper hydraulic design and the use of aeration to reduce or eliminate the parameters that trigger cavitation are extremely important (ACI 210R). Since these topics are beyond the scope of this manual, hydraulic engineers and appropriate hydraulic design manuals should be consulted.

(b) Conventional materials. While proper material selection can increase the cavitation resistance of concrete, the only totally effective solution is to reduce or eliminate the causes of cavitation. However, it is recognized that in the case of existing structures in need of repair, the reduction or elimination of cavitation may be difficult and costly. The next best solution is to replace the damaged concrete with more cavitation-resistant materials. Cavitation resistance of concrete can be increased by use of a properly designed low w/c, high-strength concrete. The use of no larger than 38-mm (1-1/2-in.) nominal maximum size aggregate is beneficial. Furthermore, methods which have reduced the unit water content of the mixture, such as use of a water-reducing admixture, are also beneficial. Vital to increased cavitation resistance are the use of hard, dense aggregate particles and a good aggregate-to-mortar bond. Typically, cement-based materials exhibit significantly lower resistance to cavitation compared to polymer-based materials.

(c) Other cavitation-resistant materials. Cavitation-damaged areas have been successfully repaired with steel-fiber concrete and polymer concrete (Houghton, Borge, and Paxton 1978). Some coatings, such as neoprene and polyurethane, have reduced cavitation damage to concrete, but since near-perfect adhesion to the concrete is critical, the use of the coatings is not common. Once a tear or a chip in the coating occurs, the entire coating is likely to be peeled off.

(d) Construction practices. Construction practices are of paramount importance when concrete surfaces are exposed to high-velocity flow, particularly if aeration devices are not incorporated in the design. Such surfaces must be as smooth as can be obtained under practical conditions. Accordingly, good construction practices as given in EM 1110-2-2000 should be followed whether the construction is new or is a repair. Formed and unformed surfaces should be carefully checked during each construction operation to confirm that they are within specified tolerances. More restrictive tolerances on surfaces should be avoided since they become highly expensive to construct and often impractical to achieve, despite the use of modern equipment and good construction practices. Where possible, transverse joints in concrete conduits or chutes should be minimized. These joints are generally in a location where the greatest problem exists in maintaining a continuously smooth hydraulic surface. One construction technique which has proven satisfactory in placement of reasonably smooth hydraulic surfaces is the traveling slipform screed. This technique can be applied to tunnel inverts and to spillway chute slabs. Hurd (1989) provides information on the slipform screed. Since surface hardness improves cavitation resistance, proper curing of these surfaces is essential.

h. Freezing and thawing.

(1) Mechanism. As the temperature of a critically saturated concrete is lowered during cold weather, the freezable water held in the capillary pores of the cement paste and aggregates expands upon freezing. If subsequent thawing is followed by refreezing, the concrete is further expanded, so that repeated cycles of freezing and thawing have a cumulative effect. By their very nature, concrete hydraulic structures are particularly vulnerable to freezing and thawing simply because there is ample opportunity for portions of these structures to become critically saturated. Concrete is especially vulnerable in areas of fluctuating water levels or under spraying conditions. Exposure in such areas as the tops of walls, piers, parapets, and slabs enhances the vulnerability of concrete to the harmful effects of repeated cycles of freezing and

thawing. The use of deicing chemicals on concrete surfaces may also accelerate damage caused by freezing and thawing and may lead to pitting and scaling. ACI 201.2R describes the action as physical. It involves the development of osmotic and hydraulic pressures during freezing, principally in the paste, similar to ordinary frost action.

(2) Symptoms. Visual examination of concrete damaged by freezing and thawing may reveal symptoms ranging from surface scaling to extensive disintegration (Figure 2-10). Laboratory examination of cores taken from structures that show surficial effects of freezing and thawing will often show a series of cracks parallel to the surface of the structure.

(3) Prevention. The following preventive measures are recommended by ACI 201.2R for concrete that will be exposed to freezing and thawing while saturated:

(a) Designing the structure to minimize the exposure to moisture. For example, providing positive drainage rather than flat surfaces whenever possible.

(b) Using a concrete with a low w/c.

(c) Using adequate entrained air to provide a satisfactory air-void system in the concrete, i.e., a bubble spacing factor of 0.20 mm (0.008 in.) or less, which will provide protection for the anticipated service conditions and aggregate size. EM 1110-2-2000 provides information on the recommended amount of entrained air.

(d) Selecting suitable materials, particularly aggregates that perform well in properly proportioned concrete.

(e) Providing adequate curing to ensure that the compressive strength of the concrete is at least 24 MPa (3,500 psi) before the concrete is allowed to freeze in a saturated state.

    i. Settlement and movement.

(1) Mechanisms.

(a) Differential movement. Situations in which the various elements of a structure are moving with respect to one another are caused by differential movements. Since concrete structures are typically very rigid, they can tolerate very little differential movement. As the differential movement increases, concrete members can be expected to be subjected to an overstressed condition. Ultimately, the members will crack or spall.

(b) Subsidence. Situations in which an entire structure is moving or a single element of a structure, such as a monolith, is moving with respect to the remainder of the structure are caused by subsidence. In these cases, the concerns are not overcracking or spalling but rather stability against overturning or sliding. Whether portions of a single structural element are moving with respect to one another or whether entire elements are moving, the underlying cause is more than likely to be a failure of the foundation material. This failure may be attributed to long-term consolidations, new loading conditions, or to a wide variety of other mechanisms. In situations in which structural movement is diagnosed as a cause of concrete deterioration, a thorough geotechnical investigation should be conducted.

(2) Symptoms. Visual examination of structures undergoing settlement or movement will usually reveal cracking or spalling or faulty alignment of structural members. Very often, movement will be apparent in nonstructural members such as block or brick masonry walls. Another good indication of structural movement is an increase in the amount of water leaking into the structure. Since differential settlement of the foundation of a structure is usually a long-term phenomenon, review of instrumentation data will be helpful in determining whether apparent movement is real. Review by structural and geotechnical engineering specialists will be required.

(3) Prevention. Prevention of settlements and movements or corrective measures are beyond the scope of this manual. Appropriate structural and geotechnical engineering manuals should be consulted for guidance.

    j. Shrinkage. Shrinkage is caused by the loss of moisture from concrete. It may be divided into two general categories: that which occurs before setting (plastic shrinkage) and that which occurs after setting (drying shrinkage). Each of these types of shrinkage is discussed in this section.

(1) Plastic shrinkage.

(a) Mechanism. During the period between placing and setting, most concrete will exhibit bleeding to some degree. Bleeding is the appearance of moisture on the surface of the concrete; it is caused by the settling of the heavier components of the mixture. Usually, the bleed water evaporates slowly from the concrete surface. If environmental conditions are such that evaporation is occurring faster than water is being supplied to the surface by bleeding, high tensile stresses can develop. These stresses can lead to the development of cracks on the concrete surface.

(b) Symptoms. Cracking caused by plastic shrinkage will be seen within a few hours of concrete placement. Typically, the cracks are isolated rather than patterned. These cracks are generally wide and shallow.

(c) Prevention. Determination of whether the weather conditions on the day of the placement are conducive to plastic shrinkage cracking is necessary. If the predicted evaporation rate is high according to ACI 305R, appropriate actions such as erecting windbreaks, erecting shade over the placement, cooling the concrete, and misting should be taken after placement. Additionally, it will be beneficial to minimize the loss of moisture from the concrete surface between placing and finishing. Finally, curing should be started as soon as is practical. If cracking caused by plastic shrinkage does occur and if it is detected early enough, revibration and refinishing of the cracked area will resolve the immediate problem of the cracks. Other measures as described above will be required to prevent additional occurrences.

(2) Drying shrinkage.

(a) Mechanism. Drying shrinkage is the long-term change in volume of concrete caused by the loss of moisture. If this shrinkage could take place without any restraint, there would be no damage to the concrete. However, the concrete in a structure is always subject to some degree of restraint by either the foundation, by another part of the structure, or by the difference in shrinkage between the concrete at the surface and that in the interior of a member. This restraint may also be attributed to purely physical conditions such as the placement of a footing on a rough foundation or to chemical bonding of new concrete to earlier placements or to both. The combination of shrinkage and restraints cause tensile stresses that can ultimately lead to cracking.

(b) Symptoms. Visual examination will typically show cracks that are characterized by their fineness and absence of any indication of movement. They are usually shallow, a few inches in depth. The crack pattern is typically orthogonal or blocky. This type of surface cracking should not be confused with thermally induced deep cracking which occurs when dimensional change is restrained in newly placed concrete by rigid foundations or by old lifts of concrete.

(c) Prevention. In general, the approach is either to reduce the tendency of the concrete, to shrink or to reduce

the restraint, or both. The following will help to reduce the tendency to shrink: use of less water in the concrete; use of larger aggregate to minimize paste content; placing the concrete at as low a temperature as practical; dampening the subgrade and the forms; dampening aggregates if they are dry and absorptive; and providing an adequate amount of reinforcement to distribute and reduce the size of cracks that do occur. Restraint can be reduced by providing adequate contraction joints.

k. Temperature changes. Changes in temperature cause a corresponding change in the volume of concrete. As was true for moisture-induced volume change (drying shrinkage), temperature-induced volume changes must be combined with restraint before damage can occur. Basically, there are three temperature change phenomena that may cause damage to concrete. First, there are the temperature changes that are generated internally by the heat of hydration of cement in large placements. Second, there are the temperature changes generated by variations in climatic conditions. Finally, there is a special case of externally generated temperature change--fire damage. Internally and externally generated temperature changes are discussed in subsequent paragraphs. Because of the infrequent nature of its occurrence in civil works structures, fire damage is not included in this manual.

(1) Internally generated temperature differences.

(a) Mechanism. The hydration of portland cement is an exothermic chemical reaction. In large volume placements, significant amounts of heat may be generated and the temperature of the concrete may be raised by more than 38 °C (100 °F) over the concrete temperature at placement. Usually, this temperature rise is not uniform throughout the mass of the concrete, and steep temperature gradients may develop. These temperature gradients give rise to a situation known as internal restraint--the outer portions of the concrete may be losing heat while the inner portions are gaining (heat). If the differential is great, cracking may occur. Simultaneously with the development of this internal restraint condition, as the concrete mass begins to cool, a reduction in volume takes place. If the reduction in volume is prevented by external conditions (such as by chemical bonding, by mechanical interlock, or by piles or dowels extending into the concrete), the concrete is externally restrained. If the strains induced by the external restraint are great enough, cracking may occur. There is increasing evidence, particularly for rehabilitation work, that relatively minor temperature differences in thin, highly restrained overlays can lead to cracking. Such cracking has been seen repeatedly in lock wall resurfacing (Figure 2-5) and in stilling basin overlays. Measured temperature differentials have typically been much below those normally associated with thermally induced cracking.

(b) Symptoms. Cracking resulting from internal restraint will be relatively shallow and isolated. Cracking resulting from external restraint will usually extend through the full section. Thermally induced cracking may be expected to be regularly spaced and perpendicular to the larger dimensions of the concrete.

(c) Prevention. An in-depth discussion of temperature and cracking predictions for massive placements can be found in ACI 207.1R and ACI 207.2R. In general, the following may be beneficial: using as low a cement content as possible; using a low-heat cement or combination of cement and pozzolans; placing the concrete at the minimum practical temperature; selecting aggregates with low moduli of elasticity and low coefficients of thermal expansion; cooling internally or insulating the placement as appropriate to minimizing temperature differentials; and minimizing the effects of stress concentrators that may instigate cracking.

(2) Externally generated temperature differences.

(a) Mechanism. The basic failure mechanism in this case is the same as that for internally generated temperature differences--the tensile strength of the concrete is exceeded. In this case the temperature change leading to the concrete volume change is caused by external factors, usually changing climatic conditions. This cause of deterioration is best described by the following examples: First, a pavement slab cast in the summer. As the air and ground temperatures drop in the fall and winter, the slab may undergo a temperature drop of 27 °C (80 °F), or more. Typical parameters for such a temperature drop (coefficient of thermal expansion of 10.8×10^{-6}/°C (6×10^{-6}/°F) indicate a 30-m (98-ft) slab would experience a shortening of more than 13 mm (1/2 in.). If the slab were restrained, such movement would certainly lead to cracking. Second, a foundation or retaining wall that is cast in the summer. In this case, as the weather cools, the concrete may cool at different rates--exposed concrete will cool faster than that insulated by soil or other backfill. The restraint provided by this differential cooling may lead to cracking if adequate contraction joints have not been provided. Third, concrete that experiences significant expansion during the warmer portions of the year. Spalling may occur if there are no adequate expansion joints. In severe cases, pavement slabs may be lifted out of alignment, resulting in so-called blowups. Fourth, concretes that have been repaired or overlayed with

materials that do not have the same coefficient of thermal expansion as the underlying material. Annual heating and cooling may lead to cracking or debonding of the two materials.

(b) Symptoms. Visual examination will show regularly spaced cracking in the case of restrained contraction. Similarly, spalling at expansion joints will be seen in the case of restrained expansion. Problems resulting from expansion-contraction caused by thermal differences will be seen as pattern cracking, individual cracking, or spalling.

(c) Prevention. The best prevention is obviously to make provision for the use of contraction and expansion joints. Providing reinforcing steel (temperature steel) will help to distribute cracks and minimize the size of those that do occur. Careful review of the properties of all repair materials will help to eliminate problems caused by temperature changes.

l. Weathering. Weathering is frequently referred to as a cause of concrete deterioration. ACI 116R defines weathering as "Changes in color, texture, strength, chemical composition, or other properties of a natural or artificial material due to the action of the weather." However, since all of these effects may be more correctly attributed to other causes of concrete deterioration described in this chapter, weathering itself is not considered to be a specific cause of deterioration.

3-3. Relating Symptoms to Causes of Distress and Deterioration

Given a detailed report of the condition of the concrete in a structure and a basic understanding of the various mechanisms that can cause concrete deterioration, the problem becomes one of relating the observations or symptoms to the underlying causes. When many of the different causes of deterioration produce the same symptoms, the task of relating symptoms to causes is more difficult than it first appears. One procedure to consider is based upon that described by Johnson (1965). This procedure is obviously idealized and makes no attempt to deal with more than one cause that may be active at any one time. Although there will usually be a combination of causes responsible for the damage detected on a structure, this procedure should provide a starting point for an analysis.

a. Evaluate structure design to determine adequacy. First consider what types of stress could have caused the observed symptoms. For example, tension will cause cracking, while compression will cause spalling. Torsion or shear will usually result in both cracking and spalling. If the basic symptom is disintegration, then overstress may be eliminated as a cause. Second, attempt to relate the probable types of stress causing the damage noted to the locations of the damage. For example, if cracking resulting from excessive tensile stress is suspected, it would not be consistent to find that type of damage in an area that is under compression. Next, if the damage seems appropriate for the location, attempt to relate the specific orientation of the damage to the stress pattern. Tension cracks should be roughly perpendicular to the line of externally induced stress. Shear usually causes failure by diagonal tension, in which the cracks will run diagonally in the concrete section. Visualizing the basic stress patterns in the structure will aid in this phase of the evaluation. If no inconsistency is encountered during this evaluation, then overstress may be the cause of the observed damage. A thorough stress analysis is warranted to confirm this finding. If an inconsistency has been detected, such as cracking in a compression zone, the next step in the procedure should be followed.

b. *Relate the symptoms to potential causes.* For this step, Table 3-3 will be of benefit. Depending upon the symptom, it may be possible to eliminate several possible causes. For example, if the symptom is disintegration or erosion, several potential causes may be eliminated by this procedure.

c. *Eliminate the readily identifiable causes.* From the list of possible causes remaining after symptoms have been related to potential causes, it may be possible to eliminate two causes very quickly since they are relatively easy to identify. The first of these is corrosion of embedded metals. It will be easy to verify whether the cracking and spalling noted are a result of corrosion. The second cause that is readily identified is accidental loading, since personnel at the structure should be able to relate the observed symptoms to a specific incident.

d. *Analyze the available clues.* If no solution has been reached at this stage, all of the evidences generated by field and laboratory investigations should be carefully reviewed. Attention should be paid to the following points:

(1) If the basic symptom is that of disintegration of the concrete surface, then essentially three possible causes remain: chemical attack, erosion, and freezing and thawing. Attempts should be made to relate the nature and type of the damage to the location in the structure and to the environment of the concrete in determining which of

the three possibilities is the most likely to be the cause of the damage.

(2) If there is evidence of swelling of the concrete, then there are two possibilities: chemical reactions and temperature changes. Destructive chemical reactions such as alkali-silica or alkali-carbonate attack that cause swelling will have been identified during the laboratory investigation. Temperature-induced swelling should be ruled out unless there is additional evidence such as spalling at joints.

(3) If the evidence is spalling and corrosion and accidental loadings have been eliminated earlier, the major causes of spalling remaining are construction errors, poor detailing, freezing and thawing, and externally generated temperature changes. Examination of the structure should have provided evidence as to the location and general nature of the spalling that will allow identification of the exact cause.

(4) If the evidence is cracking, then construction errors, shrinkage, temperature changes, settlement and movement, chemical reactions, and poor design details remain as possible causes of distress and deterioration of concrete. Each of these possibilities will have to be reviewed in light of the available laboratory and field observations to establish which is responsible.

(5) If the evidence is seepage and it has not been related to a detrimental internal chemical reaction by this time, then it is probably the result of design errors or construction errors, such as improper location or installation of a waterstop.

e. *Determine why the deterioration has occurred.* Once the basic cause or causes of the damage have been established, there remains one final requirement: to understand how the causal agent acted upon the concrete. For example, if the symptoms were cracking and spalling and the cause was corrosion of the reinforcing steel, what facilitated the corrosion? Was there chloride in the concrete? Was there inadequate cover over the reinforcing steel? Another example to consider is concrete damage caused by freezing and thawing. Did the damage occur because the concrete did not contain an adequate air-void system, or did the damage occur because the concrete used was not expected to be saturated but, for whatever reason, was saturated? Only when the cause and its mode of action are completely understood should the next step of selecting a repair material be attempted.

Table 3-3
Relating Symptoms to Causes of Distress and Deterioration of Concrete

Causes	Construction Faults	Cracking	Disintegration	Distortion/ Movement	Erosion	Joint Failures	Seepage	Spalling
Accidental Loadings		X						X
Chemical Reactions		X	X				X	
Construction Errors	X	X				X	X	X
Corrosion		X						X
Design Errors		X				X	X	X
Erosion			X		X			
Freezing and Thawing		X	X					X
Settlement and Movement		X		X		X		X
Shrinkage	X	X		X				
Temperature Changes		X				X		X

Chapter 4
Planning and Design of
Concrete Repairs

4-1. General Considerations

To achieve durable repairs it is necessary to consider the factors affecting the design and selection of repair systems as parts of a composite system. Selection of a repair material is one of the many interrelated steps; equally important are surface preparation, the method of application, construction practices, and inspection. The critical factors that largely govern the durability of concrete repairs in practice are shown in Figure 4-1. These factors must be considered in the design process so that a repair material compatible with the existing concrete substrate can be selected. Compatibility is defined as the balance of physical, chemical, and electrochemical properties and dimensions between the repair material and the concrete substrate. This balance is necessary if the repair system is to withstand all anticipated stresses induced by volume changes and chemical and electrochemical effects without distress or deterioration in a specified environment over a designated period of time. For detailed discussions of compatibility issues and the need for a rational approach to durable concrete repairs, see Emmons, Vaysburd, and McDonald (1993 and 1994).

Dimensional compatibility is one of the most critical components of concrete repair. Restrained contraction of repair materials, the restraint being provided through bond to the existing concrete substrate, significantly increases the complexity of repair projects as compared to new construction. Cracking and debonding of the repair material are often the result of restrained contractions caused by volume changes. Therefore, the specified repair material must be dimensionally compatible with the existing concrete substrate to minimize the potential for failure. Those material properties that influence dimensional

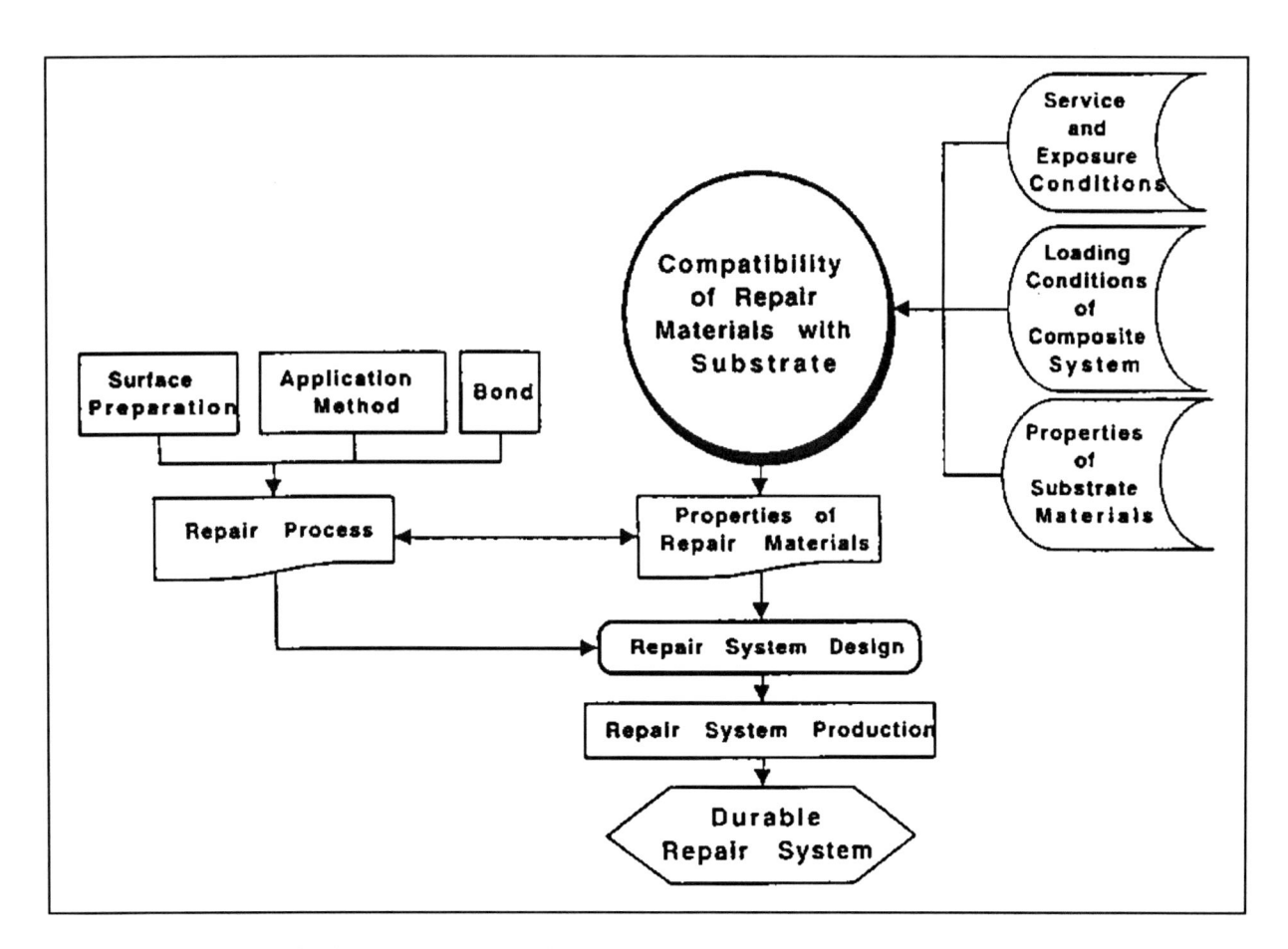

Figure 4-1. Factors affecting the durability of concrete repair systems (Emmons and Vaysburd 1995)

compatibility include drying shrinkage, thermal expansion, modulus of elasticity, and creep.

4-2. Properties of Repair Materials

In addition to conventional portland-cement concrete and mortar, there are hundreds of proprietary repair materials on the market, and new materials are continually being introduced. This wide variety of both specialty and conventional repair materials provides a greater opportunity to match material properties with specific project requirements; however, it can also increase the chances of selecting an inappropriate material. No matter how carefully a repair is made, use of the wrong material will likely lead to early repair failure (Warner 1984). Some of the material properties and their relative importance to durable repairs are discussed in the following text. These properties should be considered before any material is selected for use on a repair or rehabilitation project.

a. *Compressive strength.* Although there is some controversy over the required structural performance for many repairs, it is generally accepted that the repair material should have a compressive strength similar to that of the existing concrete substrate. Assuming the need for repair is not necessitated by inadequate strength, there is usually little advantage to be gained from repair materials with compressive strengths greater than that of the concrete substrate. In fact, significantly higher strengths of cementitious materials may indicate an excessive cement content which can contribute to higher heat of hydration and increased drying shrinkage. Repair of erosion-damaged concrete is one area in which increased strength (and corresponding higher erosion resistance) of the repair material is desirable.

b. *Modulus of elasticity.* Modulus of elasticity is a measure of stiffness with higher modulus materials exhibiting less deformation under load compared to low modulus materials. In simple engineering terms, the modulus of elasticity of a repair material should be similar to that of the concrete substrate to achieve uniform load transfer across the repaired section. A repair material with a lower modulus of elasticity will exhibit lower internal stresses thus reducing the potential for cracking and delamination of the repair.

c. *Coefficient of thermal expansion.* All materials expand and contract with changes in temperature. For a given change in temperature, the amount of expansion or contraction depends on the coefficient of thermal expansion for the material. Although the coefficient of expansion of conventional concrete will vary somewhat,

depending on the type of aggregate, it is usually assumed to be about 10.8 millionths per degree C (6 millionths per degree F). Using repair materials such as polymers, with higher coefficients of expansion, will often result in cracking, spalling, or delamination of the repair.

(1) Depending on the type of polymer, the coefficient of expansion for unfilled polymers is 6 to 14 times greater than that for concrete. Adding fillers or aggregate to polymers will improve the situation, but the coefficient of expansion for the polymer-aggregate combinations will still be one and one-half to five times that of concrete. As a result, the polymer repair material attempts to expand or contract more than the concrete substrate. This movement, when restrained through bond to the existing concrete, induces stresses that can cause cracking as the repair material attempts to contract or buckling and spalling when the repair material attempts to expand.

(2) While thermal compatibility is most important in environments that are frequently subject to large temperature changes, it should also be considered in environments in which temperature changes are not as frequent. Also, thermal compatibility is especially important in large repairs and/or overlays.

d. *Adhesion/bond.* In most cases, good bond between the repair material and the existing concrete substrate is a primary requirement for a successful repair. Bond strengths determined by slant-shear tests (ASTM C 882) are often reported by material suppliers. However, these values are highly dependent on the compressive strength of the substrate portion of the test cylinder. The test procedure requires only a minimum compressive strength of 31 MPa (4,500 psi) with no maximum strength. Therefore, these values have little or no value in comparing alternate materials unless the tests were conducted with equal substrate strengths.

(1) Bond is best specified as a surface preparation requirement. The direct tensile bond test described in ACI 503R is an excellent technique for evaluating materials, surface preparation, and placement procedures. A properly prepared, sound concrete substrate will almost always provide sufficient bond strength. In many cases, bond failures between repair materials and a properly prepared concrete substrate are a result of differential thermal strains or drying shrinkage and are not a result of inadequate bond strengths.

(2) According to ACI 503.5R, polymer adhesives provide a better bond of plastic concrete to hardened concrete than can be obtained with a cement slurry or the

plastic concrete alone. However, experience indicates that the improvement in bond is less than 25 percent as compared to properly prepared concrete surfaces without adhesives.

e. Drying shrinkage. Since most repairs are made on older structures where the concrete will exhibit minimal, if any, additional drying shrinkage, the repair material must also be essentially shrinkage-free or be able to shrink without losing bond. Shrinkage of cementitious repair materials can be reduced by using mixtures with very low w/c or by using construction procedures that minimize the shrinkage potential. Examples include dry-pack and preplaced-aggregate concrete. However, proprietary materials are being used in many repairs, often with undesirable results.

(1) A random survey of data sheets for cement-based repair materials produced in this country showed that drying shrinkage data was not even reported by some manufacturers. In those cases where data was reported, manufacturers tended to use a variety of tests and standards to evaluate the performance of their products. This arbitrary application and modification of test methods has resulted in controversy and confusion in the selection and specification of repair materials. Consequently, a study was initiated, as part of the REMR research program, to select a reliable drying shrinkage test and to develop performance criteria for selecting cement-based repair materials (Emmons and Vaysburd 1995).

(a) Three test methods are currently being evaluated in laboratory and field studies: ASTM C 157 (Modified); Shrinkage Cracking Ring; and German Angle Method. The ASTM test method, with modified curing conditions and times for length change measurements, has been used to develop preliminary performance criteria for drying shrinkage. In the modified procedure, materials are mixed and cured for 24 hr in accordance with manufacturer's recommendations. When no curing is recommended, specimens are cured in air at 50 percent relative humidity. When damp curing is recommended, specimens are placed in a moist curing room. No curing compounds are used. Following the 24 hr curing period, specimens are stored in air at 50 percent relative humidity with length change measurements at 1, 3, 28, and 60 days after casting.

(b) The ASTM C 157 (Modified) test method has been used to evaluate the drying shrinkage of 46 commercially available patching materials (Gurjar and Carter 1987). Test results at 28 days were sorted and categorized by Emmons and Vaysburd (in preparation) as shown in Figure 4-2. Shrinkage of conventional concrete

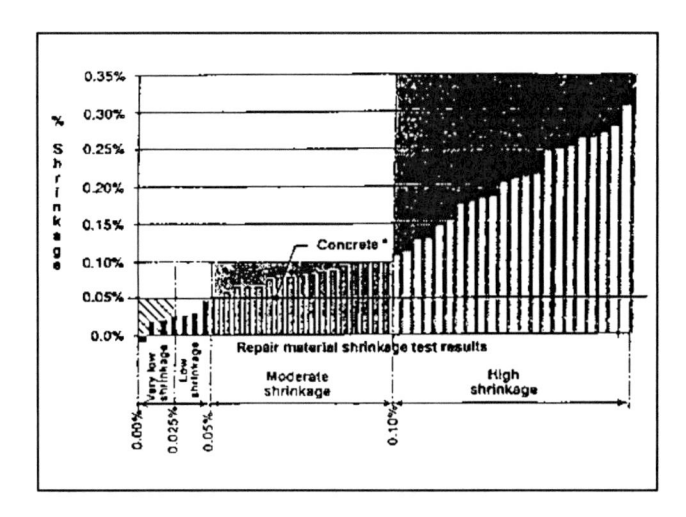

Figure 4-2. Classification of repair materials based on drying shrinkage (Emmons and Vaysburd 1995)

(0.05 percent at 28 days) was selected as a benchmark. Eighty-five percent of the materials tested had a higher shrinkage than that of concrete.

(2) Based on this work, a maximum shrinkage of 0.04 percent at 28 days (ASTM C 157 (modified)) has been proposed as preliminary performance criteria for dimensionally compatible repair materials. Final performance criteria will be selected upon completion of current large-scale laboratory and field tests to establish a correlation between laboratory test results and field performance.

f. Creep. In structural repairs, creep of the repair material should be similar to that of the concrete substrate, whereas in protective repairs higher creep can be an advantage. In the latter case, stress relaxation through tensile creep reduces the potential for cracking. It is unfortunate that most manufacturers make no mention of creep in their literature and are unable to supply basic values or to advise on environmental effects. Current tensile and compressive creep tests on selected repair materials should provide some insight into the role of creep in the overall repair system.

g. Permeability. Good quality concrete is relatively impermeable to liquids, but when moisture evaporates at a surface, replacement liquid is pulled to the evaporating surface by diffusion. If impermeable materials are used for large patches, overlays, or coatings, moisture that diffuses through the base concrete can be trapped between the substrate and the impermeable repair material. The entrapped moisture can cause failure at the bond line or critically saturate the substrate and, in the case of nonair-entrained concrete, can cause the substrate to fail if it is subjected to repeated cycles of freezing and thawing. Entrapped moisture can be a particularly troublesome problem with Corps hydraulic structures that are subject to freezing and thawing. Materials with low water absorption and high water vapor transmission characteristics are desirable for most repairs.

4-3. Application and Service Conditions

The conditions under which the repair material will be placed and the anticipated service or exposure conditions can have a major impact on design of a repair and selection of the repair material. The following factors should be considered in planning a repair strategy (Warner 1984).

a. Application conditions.

(1) Geometry. The depth and orientation of a repair section can influence selection of the repair material. In thick sections, heat generated during curing of some repair materials can result in unacceptable thermal stresses. Also, some materials shrink excessively when placed in thick layers. Some materials, particularly cementitious materials, will spall if placed in very thin layers. In contrast, some polymer-based materials can be placed in very thin sections. The maximum size of aggregate that can be used will be dictated by the minimum thickness of the repair. The repair material must be capable of adhering to the substrate without sagging when placed on vertical or overhead surfaces without forming.

(2) Temperature. Portland-cement hydration ceases at or near freezing temperatures, and latex emulsions will not coalesce to form films at temperature below about 7 °C (45 °F). Other materials may be used at temperatures well below freezing, although setting times may be increased. High temperatures will make many repair materials set faster, decrease their working life, or preclude their use entirely.

(3) Moisture. A condition peculiar to hydraulic structures is the presence of moisture or flowing water in the repair area. Generally, flowing water must be stopped by grouting, external waterproofing techniques, or drainage systems prior to repair. Some epoxy and polymer materials will not cure properly in the presence of moisture while others are moisture insensitive. Materials suitable for spall repair of wet concrete surfaces have been identified by Best and McDonald (1990a).

(4) Location. Limited access to the repair site may restrict the type of equipment, and thus the type of material that can be used for repair. Also, components of some repair materials are odorous, toxic, or combustible. Obviously, such materials should not be used in poorly ventilated areas or in areas where flammable materials aren't permitted.

b. Service conditions.

(1) Downtime. Materials with rapid strength gain characteristics that can be easily placed with minimal waste should be used when the repaired structure must be returned to service in a short period of time. Several types of rapid-hardening cements and patching materials are described in REMR Technical Note CS-MR-7.3 (USAEWES 1985g).

(2) Traffic. If the repair will be subject to heavy vehicular traffic, a high-strength material with good abrasion and skid resistance is necessary.

(3) Temperature. A material with a coefficient of thermal expansion similar to that of the concrete substrate should be used for repairs subject to wide fluctuations in temperature. High-service temperatures may adversely affect the performance of some polymer materials. Resistance to cycles of freezing and thawing will be very important in many applications.

(4) Chemical attack. Acids and sulfates will cause deterioration in cement-based materials while polymers are resistant to such chemical attack. However, strong solvents may attack some polymers. Soft water is corrosive to portland-cement materials.

(5) Appearance. If it is necessary to match the color and texture of the original concrete, many, if not most, of the available repair materials will be unsuitable. Portland-cement mixtures with materials and proportions similar to those used in the original construction are necessary where appearance is a major consideration. Procedures for repair of architectural concrete are described by Dobrowolski and Scanlon (1984).

(6) Service life. The function and remaining service life of the structure requiring repair should be considered in selection of a repair material. An extended service life requirement may dictate the choice of repair material regardless of cost. On the other hand, perhaps a lower cost, less durable, or more easily applied material can be used if the repair is only temporary.

4-4. Material Selection

Most repair projects will have unique conditions and special requirements that must be thoroughly examined before the final repair material criteria can be established. Once the criteria for a dimensionally compatible repair have been established, materials with the properties necessary to meet these criteria should be identified. A variety of repair materials have been formulated to provide a wide range of properties. Since these properties will affect the performance of a repair, selecting the correct material for a specific application requires careful study. Properties of the materials under consideration for a given repair may be obtained from manufacturer's data sheets, the REMR Repair Materials Database and *The REMR Notebook* (USAEWES 1985), evaluation reports, contact with suppliers, or by conducting tests.

a. Material properties.

(1) Manufacturer's data. Values for compressive strength, tensile strength, slant-shear bond, and modulus of elasticity are frequently reported in material data sheets provided by suppliers. However, other material properties of equal or greater importance, such as drying shrinkage, tensile bond strength, creep, absorption, and water vapor transmission, may not be reported.

(a) Experience indicates that the material properties reported in manufacturer's data sheets are generally accurate for the conditions under which they were determined. However, the designer should beware of those situations in which data on a pertinent material property is not reported. Unfavorable material characteristics are seldom reported.

(b) Material properties pertinent to a given repair should be requested from manufacturers if they are not included in the data sheets provided. General descriptions of materials, such as compatible, nonshrink, low shrinkage, etc., should be disregarded unless they are supported by data determined in accordance with standardized test methods. Material properties determined in accordance with "modified" standard tests should be viewed with caution, particularly if the modifications are not described.

(2) Repair Material Database. The REMR Repair Materials Database is described in Section 4-5. The computerized database provides rapid access to the results of tests conducted by the Corps and others; however, less than 25 percent of the available repair materials have been evaluated to date. *The REMR Notebook* currently contains 128 Material Data Sheets that include material descriptions, uses and limitations, available specifications, manufacturer's test results, and Corps test results. In addition, *The REMR Notebook* contains a number of Technical Notes that describe materials and procedures that can be used for maintenance and repair of concrete.

(3) Material suppliers. Reputable material suppliers can assist in identifying those materials and associated properties that have proven successful in previous repairs provided they are made aware of the conditions under which the material will be applied and the anticipated service conditions.

(4) Conduct tests. The formulations for commercially available materials are subject to frequent modifications for a number of reasons including changes in ownership, changes in raw materials, and new technology. Sometimes these modifications result in changes in material properties without corresponding changes to the manufacturer's data sheets or notification by the material supplier. Consequently, testing of the repair material is recommended to ensure compliance with design criteria if durability of the repair is of major importance, or the volume of repair is large (Krauss 1994).

b. Selection considerations. Concrete repair materials have been formulated to provide a wide range of properties; therefore, it is likely that more than one type of material will satisfy the design criteria for durable repair of a specific structure. In this case, other factors such as ease of application, cost, and available labor skills and equipment should be considered in selection of the repair material. To match the properties of the concrete substrate as closely as possible, portland-cement concrete or similar cementitious materials are frequently the best choices for repair. There are some obvious exceptions such as repairs that must be resistant to chemical attack. However, an arbitrary decision to repair like with like will not necessarily ensure a durable repair: The new repair material must be dimensionally compatible with the existing substrate, which has often been in place for many years.

4-5. Repair Materials Database

The Corps of Engineers Repair Material Database was developed to provide technology transfer of results from evaluations of commercial repair products performed under the REMR Research Program. The database contains manufacturer's information on uses, application procedures, limitations and technical data for approximately 1,860 commercially available repair products. In addition, Corps of Engineers test results are included for

280 products and test results from other sources for 120 products. Results of material evaluations performed by the Corps of Engineers are added to the database as reports are published. Database organization and access procedures are described in detail by Stowe and Campbell (1989) and summarized in the following.

a. Access. The database is maintained on a host computer that can be accessed by telephone via a modem using the following communication parameters:

Baud Rate:	1,200 to 9,600	Emulation:	VT-100
Data Bits:	8	Stop Bits:	1
Phone No.:	(601) 634-4223	Parity:	None

b. Operation.

(1) The database is menu driven and has help windows to facilitate its use. The products in the database are identified as either end-use or additive. An end-use product is a material that is used as purchased to make a repair, whereas an additive product is a material used in combination with other materials to produce an end-use product. The end-use products portion of the database contains products for maintenance and repair of concrete, steel, or both. The additive products portion of the database contains products that are portland-cement admixtures, binders, fibers, or special filler materials.

(2) For end-use products, product categories identify the basic type of material of which the product is composed, and for additive products, the type of end-use product for which the product is an ingredient or additive. The product uses identify the type use(s) for which the product is applicable. Keywords for searching category and use fields can be listed through the program help screens along with their definitions. Once the user selects the end-use or additive database, searches can be made by manufacturer's name, product name, product category, product use, or both category and use.

c. Assistance. For assistance or additional information regarding the database contact:

CEWES-SC-CA
3909 Halls Ferry Road
Vicksburg, MS 39180-6199
Phone: (601) 634-2814

4-6. General Categorization of Repair Approach

For ease of selecting repair methods and materials, it is helpful to divide the possible approaches into two general categories: those more suited for cracking or those more suited for spalling and disintegration. This categorization requires that some of the symptoms that were listed in Table 2-1 be regrouped as follows to facilitate selection of a repair approach:

Cracking Repair Approaches	Spalling and Disintegration Repair Approaches
Construction faults (some)	Construction faults (some)
Cracking	Disintegration
Seepage	Erosion
	Spalling

Note that distortion or movement and joint sealant failures, which were listed in Table 2-1, are not included in these categories. These are special cases that must be handled outside the process to be outlined in this chapter. Joint repair and maintenance are covered in Chapter 7. Distortion and movement are usually indications of settlement or of chemical reactions causing expansion of concrete such as severe alkali-aggregate reaction. Repairs for these conditions are beyond the scope of this manual. Materials and methods more suited for crack repairs are described in Section 4-7, while those more suited for spalling and disintegration repairs are described in Section 4-8.

4-7. Repair of Cracking

The wide variety of types of cracking described in Chapter 2 suggests that there is no single repair method that will work in all instances. A repair method that is appropriate in one instance may be ineffective or even detrimental in another. For example, if a cracked section requires tensile reinforcement or posttensioning to be able to carry imposed loads, routing and sealing the cracks with a sealer would be ineffective. On the other hand, if a concrete section has cracked because of incorrect spacing of contraction joints, filling the cracks with a high-strength material such as epoxy will only cause new cracking to occur as the concrete goes through its next contraction cycle.

a. Considerations in selecting materials and methods. Prior to the selection of the appropriate material and method for repair of cracking, the following questions should be answered (Johnson 1965):

(1) What is the nature of the cracking? Are the cracks in pattern or isolated? What is the depth of the cracking? Are the cracks open or closed? What is the extent of the cracking?

(2) What was the cause of the cracking?

(3) What was the exact mechanism of the cracking? This question requires that an analysis beyond the simple identification of the cause be conducted. For example, if the cause of the cracking has been determined to be drying shrinkage, it should then be determined whether the occurrence is the result of unusual restraint conditions or excess water content in the concrete. Understanding the mechanism will help to ensure that the same mistake is not repeated.

(4) Is the mechanism expected to remain active? Whether the causal mechanism is or is not expected to remain active will play a major role in the process to select a repair material and method.

(5) Is repair feasible? Repair of cracking caused by severe alkali-aggregate reaction may not be feasible.

(6) Should the repair be treated as spalling rather than cracking? If the damage is such that future loss of concrete mass is probable, treatment of the cracks may not be adequate. For example, cracking caused by corrosion of embedded metal or by freezing and thawing would be best treated by removal and replacement of concrete rather than by one of the crack repair methods.

(7) What will be the future movement of the crack? Is the crack active or dormant? The repair materials and techniques for active cracks are much different from the repair materials and techniques for dormant cracks. Many cracks which are still active have been "welded" together with injected epoxies only to have the crack reoccur alongside the original crack.

(8) Is strengthening across the crack required? Is the crack structural in nature? Has a structural analysis been performed as a part of the repair program?

(9) What is the moisture environment of the crack?

(10) What will be the degree of restraint for the repair material?

b. Materials and methods to consider. Once these questions have been answered, potential repair materials and methods may be selected with the procedures shown in Figures 4-3 and 4-4. All of the materials and methods listed in these figures are described in Chapter 6. In most cases, more than one material or method will be applicable. Final selection of the repair material and method must take into account the considerations discussed in Sections 4-1 through 4-4 and other pertinent project-specific conditions.

4-8. Repair of Spalling and Disintegration

Spalling and disintegration are only symptoms of many types of concrete distress. There is no single repair method that will always apply. For example, placing an air-entrained concrete over the entire surface of concrete

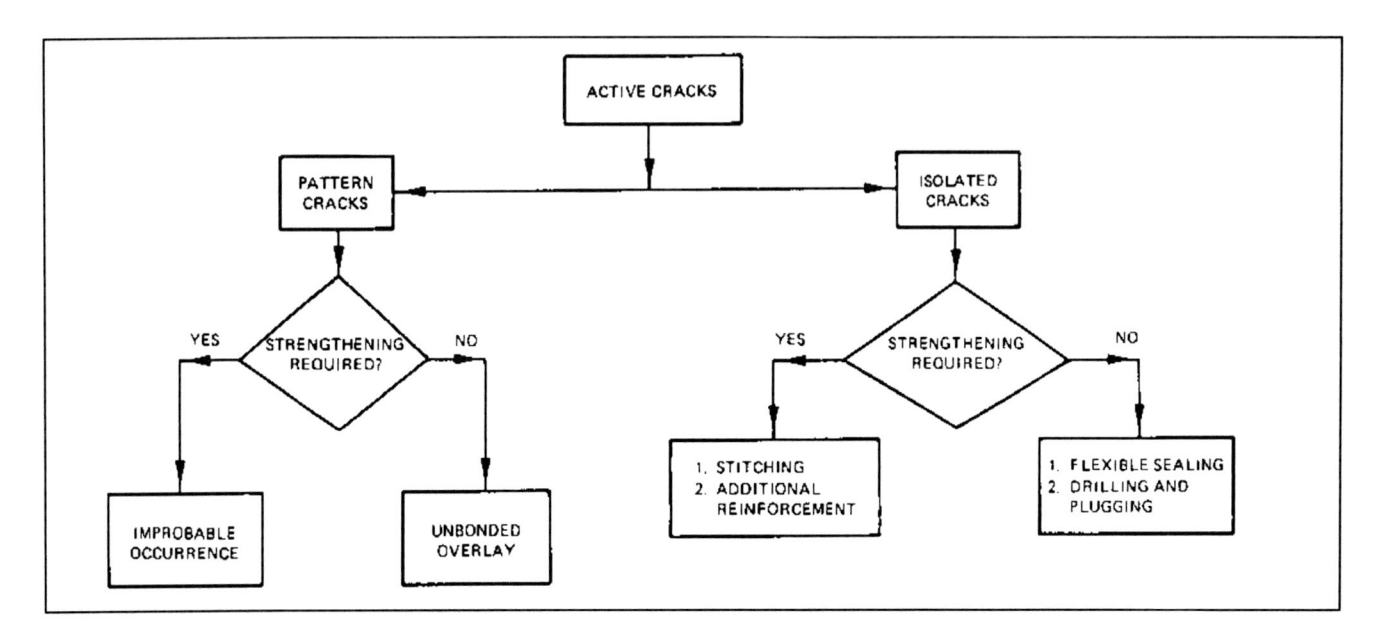

Figure 4-3. Selection of repair method for active cracks (after Johnson 1965)

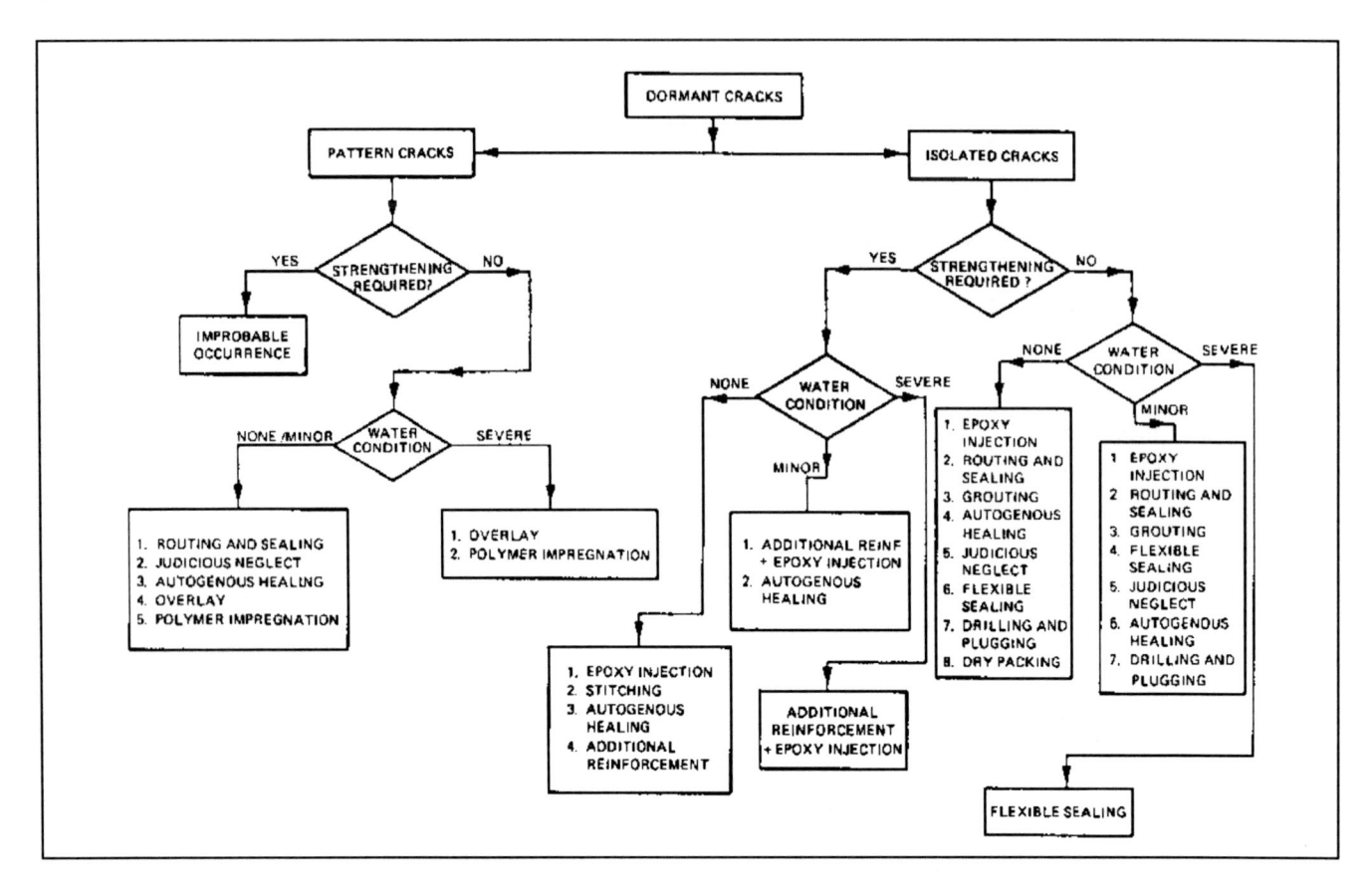

Figure 4-4. Selection of repair method for dormant cracks (after Johnson 1965)

that is deteriorating because of freezing and thawing may be a sound repair method. Use of the same technique on concrete deteriorating from strong acid attack may not be effective.

a. Considerations in selecting materials and methods. Selection of a method for repairing spalling or disintegration involves answering the following questions:

(1) What is the nature of the damage?

(2) What was the cause of the damage?

(3) Is the cause of the damage likely to remain active? If the answer to this question is yes, procedures for eliminating the factors contributing to the cause of damage should be considered. For example, if poor design details have contributed to freezing and thawing damage by allowing water to pond on a structure, drainage may be improved as part of the repair. Similarly, if attack by acid water has caused disintegration of a concrete surface, elimination of the source of the acid may eliminate acid attack as a cause of future problems. Knowledge of the future activity of a causative factor is

essential in the selection of a repair method. In the example just cited, elimination of the source of acidity might make possible a satisfactory repair with portland-cement-based material rather than a more expensive coating.

(4) What is the extent of the damage? Is the damage limited to isolated areas or is there major spalling or disintegration? The answer to this question will assist in the selection of a repair material or method that is economical and appropriate for the problem at hand.

b. General repair approach. Once these questions have been answered, a general repair approach can be selected from Table 4-1, which presents a comparison of the possible causes of spalling and disintegration symptoms and the general repair approaches that may be appropriate for each case. Table 4-2 relates the repair approaches shown in Table 4-1 to specific repair methods that are described in Chapter 6. As is true for repairing cracks, there will usually be several possible methods. The final selection must take into account the general considerations discussed in Sections 4-1 through 4-4 along with other pertinent project-specific considerations.

Table 4-1
Causes and Repair Approaches for Spalling and Disintegration

Cause	Deterioration Likely to Continue Yes	Deterioration Likely to Continue No	Repair Approach
1. Erosion (abrasion, cavitation)	X		Partial replacement Surface coatings
2. Accidental loading (impact, earthquake)		X	Partial replacement
3. Chemical reactions			
Internal	X		No action Total replacement
External	X	X	Partial replacement Surface coatings
4. Construction errors (compaction, curing, finishing)	X		Partial replacement Surface coatings No action
5. Corrosion	X		Partial replacement
6. Design errors	X	X	Partial or total replacement based on future activity
7. Temperature changes (excessive expansion caused by elevated temperature and inadequate expansion joints)	X		Redesign to include adequate joints and partial replacement
8. Freezing and thawing	X		Partial replacement No action

NOTE: This table is intended to serve as a general guide only. It should be recognized that there are probably exceptions to all of the items listed.

Table 4-2
Repair Methods for Spalling and Disintegration

Repair Approach	Repair Method
1. No action	Judicious neglect
2. Partial replacement (replacement of only damaged concrete)	Conventional concrete placement Drypacking Jacketing Preplaced-aggregate concrete Polymer impregnation Overlay Shotcrete Underwater placement High-strength concrete
3. Surface coating	Coatings Overlays
4. Total replacement of structure	Remove and replace

NOTE: Individual repair methods are discussed in Chapter 6, except those for surface coatings which are discussed in Chapter 7.

Chapter 5
Concrete Removal and Preparation for Repair

5-1. Introduction

Most repair projects involve removal of distressed or deteriorated concrete. This chapter discusses removal of concrete, preparation of concrete surfaces for further work such as overlays, preparation and replacement of reinforcing steel that has been exposed during concrete removal, and anchorage systems. Regardless of the cost or complexity of the repair method or of the material selected, the care with which deteriorated concrete is removed and with which a concrete surface is prepared will often determine whether a repair project will be successful.

5-2. Concrete Removal

a. *Alternatives.* Repair techniques requiring no concrete removal should be considered for situations where the deteriorated and damaged concrete does not threaten the integrity of the member or structure. The cost of concrete removal was saved in the rehabilitation of the tops of lock walls at Dashields Locks, U.S. Army Engineer District, Pittsburgh, by placement of an unbonded concrete overlay without removal of the deteriorated concrete. Similarly, the cost of concrete removal was saved by installation of precast concrete panels over deteriorated concrete on the backside of river walls at Lockport Lock in the U.S. Army Engineer District, Rock Island, and Troy Lock in the U.S. Army Engineer District, New York.

b. *Environment.* An evaluation to assess the impact of concrete removal debris entering a river, stream, or waterway is required before a contract is awarded. The impact varies from project to project and depends to a great extent on the size and environmental condition of the waterway and on the quantity of removal debris entering the waterway. The coarse-aggregate portion of the debris is sometimes a natural river gravel that is being returned to its place of origin and therefore its impact on the waterway is generally considered negligible. When debris fragments are of sufficient size, debris can be placed in open water to construct a fish attractor reef as an means of disposal. Recycling of concrete debris should be considered as an alternative to landfill disposal.

c. *Contract work.* If work is to be contracted, the information describing the condition and properties of the concrete must be made available at the time of invitation for bids to reduce the potential for claims by the contractor of "differing site conditions." Information provided may include type and range of deterioration, nominal maximum size and type of coarse aggregate, percentage of reinforcing steel, compressive and splitting-tensile strengths of concrete, and other pertinent information. When uncertainties exist regarding the condition of the concrete or the performance of the removal technique(s), an onsite demonstration should be implemented to test production rates and ensure acceptable results before work is begun.

d. *General considerations.* Several general considerations should be kept in mind in the selection of a concrete removal method:

(1) Usually, a repair or rehabilitation project will involve removal of deteriorated concrete. However, for many maintenance and repair projects, concrete is removed to a fixed depth to ensure that the bulk of deteriorated concrete is removed or to accommodate a specific repair technique. For some projects, this requirement would cause a significant amount of sound concrete to be removed and, thereby, a change in removal method(s), since some methods are more cost effective for sound concrete than others.

(2) Selected concrete removal methods should be safe and economical and should have as little effect as possible on concrete remaining in place. Selection of a proper removal method may have a significant effect on the length of time that a structure must be out of service. Some methods permit a significant portion of the work to be accomplished without removing the structure from service. For example, drilling of boreholes in a lock wall in conjunction with removal of concrete by blasting may be done while the lock is operational.

(3) The same removal method may not be suited for all portions of a given structure. The most appropriate method for each portion of the structure should be selected and specified.

(4) More than one removal method may be required for a particular area. For example, a presplitting method may be used to fracture and weaken the concrete to be removed, while an impacting method is used to complete the removal for the same location.

(5) In some instances, a combination of removal methods may be used to limit damage to concrete that is not being removed. For example, a cutting method may

be used to delineate an area in which an impacting method is to be used as the primary means of removal.

(6) Field tests of various removal methods are very well suited for demonstration projects done during the design phase of a major repair or rehabilitation project.

(7) The cost of removal and repair should be compared to the cost of total demolition and replacement of the member or structure if the damage is extensive.

(8) Care should be taken to avoid embedded items such as electrical conduits and gate anchorage's. Dimensions and locations of embedded items documented in the as-built drawings should not be taken for granted.

e. Classification of concrete removal methods. Removal methods may be categorized by the way in which the process acts on the concrete. These categories are blasting, crushing, cutting, impacting, milling, and presplitting. Table 5-1 provides a general description of these categories and lists the specific removal methods within each category. Table 5-2 provides a summary of information on each method. These methods are discussed in detail in the following. See Campbell (1982) for additional information.

f. Blasting methods. Blasting methods employ rapidly expanding gas confined within a series of boreholes to produce controlled fracture and removal of concrete (Figure-5-1). Explosive blasting, the only blasting method commercially available in the United States, is applicable for concrete removal from mass concrete structures where 250 mm (10 in.) or more of face is to be removed and the volume of removal is significant. Explosive blasting is considered to be the most expedient and, in many cases, the most cost-effective means of removal from mass concrete structures. Its primary disadvantage is its potential for damage to the remaining concrete and adjacent structures. Blasting plans typically include drilling holes along removal boundary and employing controlled and sequential blasting methods for the removal. A commonly employed, controlled blasting technique, smooth blasting, uses detonating cord to distribute the blast energy throughout the hole, thereby, avoiding energy concentrations that might damage the concrete that remains. Cushion blasting, a more protective but less used control, is the same as smooth blasting except wet sand is used to fill holes and cushion against the blast effect. The use of saw cuts along removal perimeters is recommended to reduce overbreakage. For removal of vertical faces, a full-depth cut is recommended along the bottom boundary. Sequential blasting techniques allow

more delays to be employed per firing. They are recommended for optimizing the amount of explosive detonated per firing while maintaining air-blast pressures, ground vibrations, and fly rock at acceptable levels. When uncertainties regarding the blast plan exist, a pilot test program is recommended to evaluate parameters and ensure acceptable results. Because of dangers inherent in handling and using explosives, all phases of the blasting project should be performed and monitored for compliance with EM 385-1-1.

g. Crushing methods. Crushing methods employ hydraulically powered jaws to crush and remove the concrete.

(1) Boom-mounted mechanical crushers. Boom-mounted crushers (Figure 5-2) are applicable for removing concrete from decks, walls, columns, and other concrete members where the shearing plane depth is 1.8 m (6 ft) or less. This method is typically more applicable for total demolition of a member(s) than for partial removal for rehabilitation or repair. Pulverizing jaw attachments that crush and debond the concrete from the reinforcing steel to facilitate their separation for recycling are available. The major limitations are that the removal boundary must be saw cut to reduce overbreakage, crushing must be started from a free edge or hole made by hand-held breakers or other means, and the exposed reinforcing is damaged beyond reuse. Care must be taken to avoid damaging members that are to support the repair.

(2) Portable mechanical crushers. Portable crushers are applicable for removing concrete from decks, walls, columns, and other concrete members where the shearing plane depth is 300 mm (12 in.) or less. The crusher weighs approximately 45 kg (100 lb) and requires two men to handle. The major limitations are that the removal boundary must be saw cut to reduce overbreakage, crushing must be started from a free edge or hole made by hand-held breakers or other means, and the exposed reinforcing is damaged beyond reuse.

h. Cutting methods. Cutting methods employ full depth perimeter cuts to disjoint concrete for removal as a unit(s). The maximum size of the unit(s) is determined by the load carrying capacities of available lifting and transporting equipment. Cutting methods include abrasive water jets, diamond saws, stitch drilling, and thermal tools.

(1) Abrasive-water-jet cutting. Water-jet systems that include abrasives are applicable for making cutouts through slabs, walls, and other concrete members where

Table 5-1
A General Classification of Concrete Removal Methods Applicable for Concrete Repair

Category	Description	Specific Methods
Blasting	Blasting methods employ rapidly expanding gas confined within a series of boreholes to produce controlled fracture and removal of concrete	Explosive blasting
Crushing	Crushing methods employ hydraulically powered jaws to crush and remove the concrete	Mechanical crushing, boom-mounted Mechanical crushing, portable
Cutting	Cutting methods employ full-depth perimeter cuts to disjoint concrete for removal as a unit or units	Abrasive-water-jet cutting Diamond-blade cutting Diamond-wire cutting Stitch drilling Thermal cutting
Impacting	Impacting methods employ repeated striking of the surface with a mass to fracture and spall the concrete	Mechanical impacting, hand-held Mechanical impacting, boom-mounted Mechanical impacting, spring-action
Milling	Milling methods generally employ abrasion or cavitation-erosion techniques to remove concrete from surfaces	Hydromilling Rotary head milling
Presplitting	Presplitting methods employ wedging forces in a designed pattern of boreholes to produce a controlled cracking of the concrete to facilitate removal of concrete by other means	Presplitting, chemical-expansive agents Presplitting, piston-jack splitter Presplitting, plug-and-feather splitter

Table 5-2
Selection Features and Considerations for Concrete Removal Methods

Category	Method	Features	Considerations
Blasting	Explosive blasting	Method applicable for removal from mass concrete structures	Requires highly skilled personnel for design and execution of blasting plan
		Method is most expedient and, in many cases, the most cost-effective means of removing large volumes where 250 mm (10 in.) or more of face is to be removed	Stringent safety regulations must be complied with regard to the transportation, storage, and use of explosives because of their inherent dangers
		Produces reasonably small size debris that is easily handled	Sequential blasting techniques must be employed to reduce peak blast energies and, thereby, limit damage to surrounding property resulting from air-blast pressure, ground vibration, and fly rock
			Control blasting techniques should be employed to limit damage to concrete that remains
Crushing	Mechanical crushing, boom-mounted	Method applicable for removing concrete from decks, walls, columns, and other concrete members where shearing plane depth is 1.8 m (6 ft) or less	Method is more applicable for total demolition of a concrete member than for removal to rehabilitate or repair
		Boom allows removal from vertical and overhead members	Boundaries must be saw cut to limit overbreakage
		Steel reinforcing can be cut	Removal must be started from a free edge or a hole cut in member
		Limited noise and vibration is produced	Exposed reinforcing steel is damaged beyond reuse
		Pulverizing jaw attachment can debond the concrete from the steel reinforcement for purpose of recycling both	Production rates vary depending on condition of concrete and amount of reinforcement
		Method produces relatively small debris that is easily handled	
	Mechanical crushing, portable	Method applicable for removal from decks, walls, and other members where shearing plane depth is 300 mm (12 in.) or less	Requires two men to handle (weighs approximately 45 kg (100 lb))
			Reinforcing steel is damaged beyond reuse
		Method can be used to remove concrete in areas of limited work space	Crushing must be started from a free edge or a hole cut in member
			Boundaries must be saw cut to limit overbreakage

Table 5-2 (Continued)

Category	Method	Features	Considerations
		Limited noise and vibration is produced	Production rates are low
		Produces small size debris that is easily handled	
Cutting	Abrasive-water-jet cutting	Method applicable for making cutouts through slabs, walls, and other concrete members where access to only one face is feasible and depth of cut is 500 mm (20 in.) or less	Cutting is typically slower and more costly than diamond-blade sawing
			Controlling flow of waste water may be required
		Abrasives enable jet to cut steel reinforcing and hard aggregates	Personnel must wear hearing protection because of the high levels of noise produced
		Irregular and curved cutouts can be made	Additional safety precautions are required because of high water pressures (200 - 340 MPa (30,000 - 50,000 psi)) produced by system
		Cutouts can be made without overcutting corners	
		Cuts can be made flush with adjoining members	
		No heat, vibration, or dust is produced	
		Handling of debris is more efficient as bulk of concrete is removed as units	
	Diamond-blade cutting	Method applicable for making cutouts through slabs, walls, and other concrete members where access to only one face is feasible and depth of cut is 600 mm (24 in.) or less	Selection of the type diamonds and metal bond used in blade segments is based on the type (hardness) and percent of coarse aggregate and on the percent of steel reinforcing in cut
			The higher the percent of steel reinforcement in cuts, the slower and more costly the cutting
		Precision cuts can be made	The harder the aggregate, the slower and more costly the cutting
		No dust or vibration is produced	Controlling flow of waste water may be required
		Handling of debris is more efficient as bulk of concrete is removed as units	Special blades with flush-cut arbors are required to make cuts flush with adjoining members

Table 5-2 (Continued)

Category	Method	Features	Considerations
	Diamond-wire cutting	Method applicable for making cutouts through concrete where depth of cut is greater than can be economically cut with the diamond-blade saw	The wire saw is a specialty tool that for many jobs will not be as cost effective as other techniques, such as blasting, impacting, and presplitting.
		Cuts can be made through mass concrete and in areas of difficult access	Selection of type diamonds and metal bond used in beads is based on type (hardness) and percent of coarse aggregate and percent of steel reinforcing in cut
		Overcutting of corner can be avoided if cut started from drilled hole at corner	The higher the percent of steel reinforcement in cuts, the slower and more costly the cutting
		No dust or vibration is produced	The harder the aggregate, the slower and more costly the cutting
		Handling of debris is more efficient as bulk of concrete is removed as units	Beads with embedded diamonds last longer, but are more expensive than beads with electroplated diamonds (single layer)
			Wires with beads having embedded diamonds should be of sufficient length to complete cut as replacement will not fit into cut (wear reduces wire diameter and, thereby, cut opening as cutting proceeds).
			Deep cutouts that are formed by three or more boundary cuts may require tapering to avoid binding during removal
			Controlling flow of waste water may be required
	Stitch drilling	Method applicable for making cutouts through concrete members where access to only one face is feasible and depth of cut is greater than can be economically cut by diamond-blade saw	Rotary-percussion drilling is significantly more expedient and economical than diamond-core for nonreinforced concrete
			Diamond-core drilling is more applicable than rotary-percussion drilling for reinforced concrete
		Handling of debris is more efficient as bulk of concrete is removed as units	The greater the percentage of steel reinforcement contained within a cut, the slower and more costly is the cutting
			Depth of cuts is dependent on accuracy of drilling equipment in maintaining overlap between holes with depth and on the diameter of boreholes drilled
			The deeper the cut, the greater borehole diameter required to maintain overlap between adjacent holes and the greater the cost
			Uncut portions between adjacent boreholes will prevent removal

(Sheet 3 of 8)

Table 5-2 (Continued)

Category	Method	Features	Considerations
			Concrete toughness for percussion drilling and aggregate hardness for diamond coring will affect cutting rate and cost
			Personnel must wear hearing protection because of the high levels of noise produced
	Thermal cutting	Method applicable for making cutouts through heavily reinforced decks, beams, walls, and other reinforced members where site conditions allow efficient flow of molten concrete from cuts	Method is of limited commercial availability and is costly
			Remaining concrete has thermal damage with more extensive damage occurring around steel reinforcement
		Method is an effective means of cutting prestressed members	Noise, smoke, and fumes are produced
			Personnel must be protected from heat and hot flying rock produced by cutting operation
		Irregular shapes can be cut	
		Minimal vibration and dust produced	Additional safety precautions are required because of hazards associated with storage, handling, and use of compressed and flammable gases
		Handling of debris is more efficient as bulk of concrete is removed as units	
Impacting	Mechanical impacting, boom-mounted breaker	Method is applicable for both full and partial depth removals where required production rates are greater than can be economically achieved by the use of hand-held breakers	The blow energy delivered to the concrete should be limited to protect the structure being repaired and surrounding structures from damage resulting from the high cyclic energy generated
			Performance is function of concrete soundness and toughness
		Boom allows concrete to be removed from vertical and overhead members	Productivity is significantly reduced when boom is operated from top of wall because of the operator's limited view of the removal operation
		Boom-mounted breakers are widely available commercially	Care must be taken to avoid damage to supporting members
		Method produces easily handled debris	Concrete that remains may be damaged (microcracking) along with reinforcing steel
			Saw cuts at boundaries should be employed to reduce the occurrence of feathered edges
			Dust is produced
			Personnel must wear hearing protection because of the high levels of noise produced

(Sheet 4 of 8)

Table 5-2 (Continued)

Category	Method	Features	Considerations
	Mechanical impacting, hand-held breaker	Method is applicable for work involving limited volumes of concrete removal and for removal in areas of limited access	Hand-held breakers are generally not applicable for large volumes of removal, except where blow energy must be limited
			Performance is function of concrete soundness and toughness
		Hand-held breakers are widely available commercially	Significant loss in productivity occurs when breaking action is other than downward
		Breakers can be operated by unskilled labor	Removal boundaries will likely require 25-mm (1-in.) deep or greater saw cut to reduce the occurrence of feathered edges
		Method produces relatively small debris that is easily handled	Concrete that remains may be damaged (microcracking)
			Size of breakers for bridge decks is typically limited to 14-kg (30-lb) class for removal above reinforcement and 7-kg (15-lb) class from around reinforcement
			Dust is produced
			Personnel must wear hearing protection because of the high levels of noise produced
	Mechanical impacting, spring-action hammer	Method is applicable for breaking concrete pavement, decks, walls, and other thin members where production rates required are greater than can be economically achieved by the use of hand-held breakers	Method is more applicable for total demolition of a concrete member than for removal to rehabilitate or repair
			The blow energy delivered to the concrete should be limited to protect the structure being repaired and surrounding structures from damage resulting from the high cyclic energy generated
		For decks, hammer can completely punch through slab with each blow leaving only the reinforcing steel	Care must be taken to avoid damage to supporting members
			Performance is function of concrete soundness and toughness
		Method produces easily handled debris	Concrete that remains may be damaged (microcracking) along with reinforcing steel
			Saw cuts at boundaries should be employed to reduce the occurrence of feathered edges
Milling	Hydromilling (Also known as hydrodemolition and water-jet blasting)	Method is applicable for removal of deteriorated concrete from surfaces of decks and walls where removal depth is 150 mm (6 in.) or less	Method is costly
			Productivity is significantly reduced when sound concrete is being removed

Table 5-2 (Continued)

Category	Method	Features	Considerations
		Method does not damage the concrete that remains	Removal profile will vary with changes in depth of deterioration
			Holes through member (blowouts) are a common occurrence when removal is near full depth of member
		Steel reinforcing is left undamaged for reuse	
			Repair of blowouts requires additional material and form work, thereby, increasing repair time and cost
		Method produces easily handled, aggregate-size debris	
			Method requires large source of potable water (the water demand for some units exceeds 4,000 L/hr (1,000 gal/hr))
			Laitence coating that is deposited on remaining surfaces during removal should be washed from surface before coating dries
			Flow of waste water may have to be controlled
			An environmental impact statement will be required if waste water is to enter a waterway
			Personnel must wear hearing protection because of the high level of noise produced
			Fly rock is produced
			Additional safety requirements are required because of the high pressures (100 - 300-MPa (16,000 - 40,000-psi) range) produced by the system
	Rotary-head milling	Method is applicable for removing deteriorated concrete from mass structures	Removal is limited to concrete outside structural steel reinforcement
		Method is applicable for removing deteriorated concrete cover from reinforced members such as pavements and decks where it is unlikely that the reinforcement will be contacted	Significant loss of productivity occurs in sound concrete
			Productivity is significantly reduced when boom is operated from top of wall as operator's view of cutting is very limited
			Concrete that remains may be damaged (microcracking)
		Boom allows removal from vertical and overhead surfaces	Skid loader units typically mill a more uniform removal profile than other rotary-head and water-jet units

Table 5-2 (Continued)

Category	Method	Features	Considerations
		Concrete containing wire mesh can be cut without significant losses in productivity	Noise, vibration, and dust are produced
		Method produces relatively small debris that is easily handled	
Presplitting	Chemical presplitting, expansive agents	Method is applicable for presplitting concrete members where depth of boreholes is 10 times borehole diameter or greater	Personnel must be restricted from presplitting area during early hours of product hydration as material has the potential to blow out of boreholes and cause injury
		Expansive products can be used to produce vertical presplitting planes of significant depth	Presplitting with expansive agents is typically costly
			Expansive products that are prills or become slurries when water is added are best used in gravity-filled, vertical, or near-vertical holes. A liner may be required to contain the expansive material in holes drilled into concrete with extensive cracks
		Some products form a clay-type material when mixed with water that allows the material to be packed into horizontal holes	Products are limited to a specific temperature range
			Rotary-head milling or mechanical-impacting methods will be required to complete removal
		No vibration, noise, or flying rock is produced other than that produced by the drilling of boreholes and the secondary breakage method	Development of presplitting plane is significantly decreased by presence of reinforcing steel normal to plane
			Loss of control of presplitting plane can result if boreholes are too far apart or holes are located in severely deteriorated concrete
	Mechanical presplitting, piston-jack splitter	Method is applicable for presplitting more massive concrete structures where 250 mm (10 in.) or more of face is to be removed and presplitting requires boreholes of a depth greater than can be used by plug-and-feather splitters	Large-diameter (90-mm (3-1/2-in.)) boreholes are required that increase cost
			Splitters are typically used in pairs to control presplitting plane
		Splitter can be reinserted into boreholes to continue removal for full depth of holes	Hand-held breakers and pry bars are typically required to complete removal
			Development of presplitting plane is significantly decreased by presence of reinforcing steel normal to presplit plane
		Splitter can be used in areas of difficult access	Loss of control of presplitting plane can result if boreholes are too far apart or holes are located in severely deteriorated concrete

Table 5-2 (Concluded)

Category	Method	Features	Considerations
			Availability of splitters is limited in the U.S.
		No vibration, noise, or flying rock is produced other than that produced by the drilling of boreholes and the secondary breakage method	
	Mechanical presplitting, plug-and-feather splitter	Method applicable for presplitting slabs, walls, and other concrete members where presplitting depth is 4 ft or less	Splitter can not be reinserted into boreholes to continue presplitting after presplit section has been removed, as the body of the tool is wider than the borehole
		Method typically less costly than cutting methods	Development of presplitting plane in direction of borehole depth is limited
		Initiation of direction of presplitting can be controlled by orientation of plug and feathers	Development of presplitting plane is significantly decreased by presences of reinforcing steel normal to plane
		Splitters can be used in areas of limited access	Secondary means of breakage will typically be required to complete removal
		Limited skills required by operator	Loss of control of presplitting plane can result if boreholes are too far apart or holes are located in severely deteriorated concrete
		No vibration, noise, or flying rock is produced other than that produced by the drilling of boreholes and the secondary breakage method	

(Sheet 8 of 8)

Figure 5-1. Surface removal of deteriorated concrete by explosive blasting

Figure 5-2. Boom-mounted concrete crusher

access to only one face is feasible and depth of cut is 500 mm (20 in.) or less. The abrasives enable the jet to cut steel reinforcing and hard aggregates. One major limitation of abrasive-water-jet cutting is that it is typically slower and more costly than diamond-blade sawing. Personnel must wear hearing protection because of the high levels of noise produced. Additional safety precautions are required because of high water pressures (200 to 340 MPa (30,000 to 50,000 psi)) produced by the system. Controlling flow of waste water may be required.

(2) Diamond-blade cutting. Diamond-blade cutting (Figure 5-3) is applicable for making cutouts through slabs, walls, and other concrete members where access to only one face is feasible and depth of cut is 600 mm (24 in.) or less. Blade selection is a function of the type (hardness) and percent of coarse aggregate and on the percent of steel reinforcing. The harder the coarse aggregate and the higher the percentage of steel reinforcement in the cut, the slower and more costly the cutting. Diamond-blade cutting is also applicable for making cuts along removal boundaries to reduce feathered edges in support of other methods.

(3) Diamond-wire cutting. Diamond-wire cutting (Figure 5-4) is applicable for making cutouts through concrete where the depth of cut is greater than can be economically cut with a diamond-blade saw. Cuts can be made through mass concrete and in areas of difficult access. The cutting wire is a continuous loop of multi-strand wire cable strung with steel beads containing either embedded or electroplated diamonds. Beads with embedded diamonds last longer but are more expensive than beads with electroplated diamonds (single layer). Wires with beads having embedded diamonds should be of sufficient length to complete the cut as replacement wire will not fit into the cut (wear reduces wire diameter and, thereby, cut opening as cutting proceeds). The wire saw is a specialty tool that for many jobs will not be as

Figure 5-3. Diamond-blade saw

Figure 5-4. Diamond-wire saw

cost effective as other methods, such as blasting, impacting, and presplitting.

(4) Stitch cutting. Method applicable for making cutouts through concrete members where access to only one face is feasible and depth of cut is greater than can be economically cut by diamond-blade saw. Depth of cuts is dependent on the accuracy of drilling equipment in maintaining overlap between holes with depth and on the diameter of boreholes drilled. If overlap between holes is not maintained, uncut portions of concrete that will prevent removal remain between adjacent boreholes. If opposite faces of a member can be accessed, diamond-wire cutting will likely be more applicable. Concrete toughness for percussion drilling and aggregate hardness for diamond coring will affect the cutting rate and the cost.

(5) Thermal cutting. Thermal-cutting methods are applicable for making cutouts through heavily reinforced decks, beams, walls, and other reinforced members where site conditions allow efficient flow of molten concrete from cuts. Flame tools (Figure 5-5) are typically employed for cutting depths of 600 mm (24 in.) or less, and lances (Figure 5-6), for greater depths. Thermal cutting tools are of limited commercial availability and are costly to use. The concrete that remains has a layer of thermal damage with more extensive damage occurring around steel reinforcement. Personnel must be protected from heat and hot flying rock produced by the cutting operation. Additional safety precautions are required because of the hazards associated with the storage, handling, and use of compressed and flammable gases. The

Figure 5-5. Powder torch

Figure 5-6. Thermal lance

method is also applicable for the demolition of prestressed members.

i. Impacting methods. Impacting methods generally employ the repeated striking of a concrete surface with a mass to fracture and spall the concrete. Impact methods are sometimes used in a manner similar to cutting methods to disjoint the concrete for removal as a unit(s) by breaking out concrete along the removal perimeter of thin members such as slabs, pavements, decks, and walls. Any reinforcing steel along the perimeter would have to be cut to complete the disjointment. Impacting methods include the boom-mounted and hand-held breakers and spring-action hammers.

(1) Boom-mounted breakers. Boom-mounted impact breakers are applicable for both full- and partial-depth removals where production rates required are greater than can be economically achieved by the use of hand-held breakers. The boom-mounted breakers are somewhat similar to the hand-held breakers except that they are considerably more massive. The tool is normally attached to the hydraulically operated arm of a backhoe or excavator (Figure 5-7) and can be operated by compressed air or hydraulic pressure. The reach of the hydraulic arm enables the tool to be used on walls at a considerable distance above or below the level of the machine. Boom-mounted breakers are a highly productive means of removing concrete. However, the blow energy delivered to the concrete should be limited to protect the structure being repaired and surrounding structures from damage resulting from the high cyclic energy generated. Saw cuts

should be employed at removal boundaries to reduce the occurrence of feathered edges. The concrete that remains may be damaged (microcracking) along with the exposed reinforcing steel. Washing the concrete surface with a high-pressure (138 MPa (20,000 psi) minimum) water jet may remove some of the microfractured concrete.

(2) Spring-action hammers. Spring-action hammers (sometimes referred to as mechanical sledgehammers) are boom-mounted tools that are applicable for breaking concrete pavements, decks, walls, and other thin members where production rates required are greater than can be economically achieved with the use of hand-held breakers. Hammers are more applicable for total demolition of a concrete member than for removal to rehabilitate or repair. The arm of the hammer is hydraulically powered, and the impact head is spring powered. The spring is compressed by the downward movement of the arm of the backhoe or excavator and its energy released just prior to impact. There are truck units available that make it easier to move between projects. The operation of the hammer and advancement of the truck during removal are controlled from a cab at the rear of truck (Figure 5-8). The blow energy delivered to the concrete should be limited to protect the structure being repaired and surrounding structures from damage caused by the high cyclic energy generated. Saw cuts should be employed at removal boundaries to reduce the occurrence of feathered edges. The concrete that remains may be damaged (microcracking) along with the exposed reinforcing steel.

Figure 5-8. Spring-action hammer (mechanical sledgehammer)

Figure 5-7. Boom-mounted breaker

(3) Hand-held impact breakers. Hand-held impact breakers (Figure 5-9) are applicable for work involving limited volumes of concrete removal and for removal in areas of limited access. Hand-held breakers are sometimes applicable for large volumes of removal where blow energy must be limited or the concrete is highly deteriorated. Breakers are also suitable for use in support of other means of removal. Hand-held breakers are powered by one of four means: compressed air, hydraulic pressure, self-contained gasoline engine, or self-contained electric motor.

j. Milling. Milling methods generally employ impact-abrasion or cavitation-erosion techniques to remove concrete from surfaces. Methods include hydromilling and rotary-head milling.

(1) Hydromilling. Hydromilling (also known as hydrodemolition and water-jet blasting) is applicable for removal of deteriorated concrete from surfaces of decks (Figure 5-10) and walls where removal depth is 150 mm (6 in.) or less. This method does not damage the concrete that remains and leaves the steel reinforcing undamaged for reuse in the replacement concrete. Its major limitations are that the method is costly, productivity is significantly reduced when sound concrete is being removed, and the removal profile varies with changes in depth of deterioration. Holes through members (blowouts) are a common occurrence when removal is near full depth of a member. This method requires a large source of potable water (the water demand for some units exceeds 4,000 L/hr (1,000 gal/hr)). An environmental impact statement is required if waste water is to enter a waterway. Personnel must wear hearing protection because of

Figure 5-10. Hydromilling (water-jet blasting)

the high level of noise produced. Flying rock is produced. Laitence coating that is deposited on remaining surfaces during removal should be washed from the surfaces before the coating dries.

(2) Rotary-head milling. Method is applicable for removing deteriorated concrete from mass structures (Figure 5-11) and for removing deteriorated concrete cover from reinforced members such as pavements and decks where its contact with the reinforcement is unlikely. Removal is limited to concrete outside structural steel reinforcement. Significant loss of productivity occurs in sound concrete. For concrete having a compressive strength of 55 MPa (8,000 psi) or greater, rotary-head milling is not applicable. Concrete that remains may be

Figure 5-9. Hand-held breaker

Figure 5-11. Rotary-head milling

damaged (microcracking). Skid loader units typically mill a more uniform removal profile than other rotary-head and water-jet units.

    k. Presplitting. Presplitting methods employ wedging forces in a designed pattern of boreholes to produce a controlled cracking of the concrete to facilitate removal of concrete by other means. The pattern, spacing, and depth of the boreholes affect the direction and extent of the presplitting planes. Presplitting methods include chemical-expansive agents and hydraulic splitters. Note: for all presplitting methods, the development of a presplitting plane is significantly decreased by the presence of reinforcing steel normal to the plane, and the loss of control of a presplitting plane can result if boreholes are too far apart or holes are located in severely deteriorated concrete.

    (1) Chemical presplitting, expansive agents. The presplitting method that uses chemical-expansive agents (Figure 5-12) is applicable for removal from slabs, walls, and other concrete members where depth of boreholes is 10 times the borehole diameter or greater. It is especially applicable for situations requiring the development of vertical presplitting planes of significant depth. The main disadvantages of employing expansive agents are cost and application-temperature limitations. Personnel must be restricted from the presplitting area during early hours of product hydration as the material has the potential to blow out of boreholes and cause injury. Expansive products that are prills or become slurries when water is added are best used in gravity filled, vertical or near-vertical holes. Some products form a clay-type material when mixed with water that allows the material to be packed into

horizontal holes. The newer expansive agents produce presplitting planes in 4 hr or less. Rotary-head milling or mechanical-impacting methods will be required to complete removal.

    (2) Mechanical presplitting, piston-jack splitter. Piston-jack splitters (Figure 5-13) are applicable for presplitting more massive concrete structures where 250 mm (10 in.) or more of the face is to be removed and presplitting requires boreholes of a depth greater than can be used by plug-and-feather splitters. The piston-jack splitters initiate presplitting from opposite sides of a borehole, normal to the direction of piston movement. The splitters are reinserted into boreholes to continue removal. Process is repeated for full depth of holes. Splitters are typically used in pairs to control the presplitting plane. The primary disadvantages of this method are the cost of drilling the required 90-mm (3-1/2-in.)-diam boreholes and the limited availability of piston-jack devices in the United States.

    (3) Mechanical presplitting, plug-feather splitter. Plug-and-feather splitters (Figure 5-14) are applicable for presplitting slabs, walls, and other concrete members where the presplitting depth is 1.2 m (4 ft) or less. Initiation of direction of presplitting can be controlled by orientation of plug and feathers. The primary limitation of these splitters is that they can not be reinserted into boreholes to continue presplitting after the presplit section has been removed, since the body of the tool is wider than the borehole.

    l. Monitoring removal operations. The extent of damage to the concrete that remains after a removal

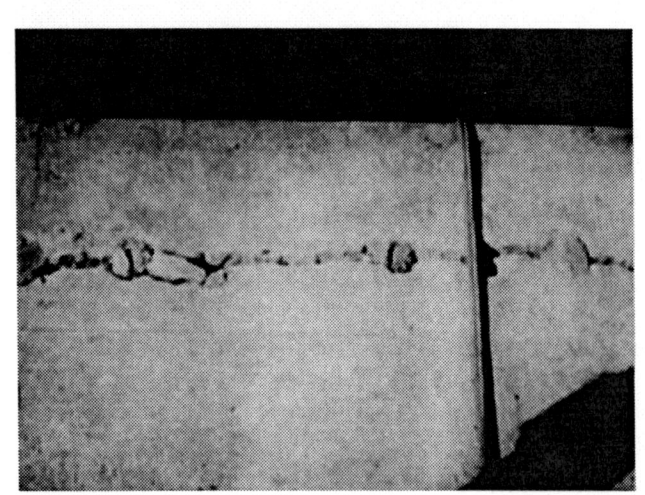

Figure 5-12. Presplitting using chemical-expansive agent

Figure 5-13. Piston-jack splitter

method has been employed is usually evaluated by visual inspection of the remaining surfaces. For a more detailed evaluation, a monitoring program can be implemented. The program may consist of taking cores before and after removal operations, making visual and petrographic examinations, and conducting pulse-velocity and ultimate-strength tests of the cores. A pulse-velocity study of the in situ concrete may also be desired. A comparison of the data obtained before and after removal operations could then be used to determine the relative condition of remaining concrete and to identify damage resulting from the removal method employed. To further document the extent of damage, an instrumentation program may be required.

m. Quantity of concrete removal. In most concrete repair projects, all damaged or deteriorated concrete should be removed. However, estimating the quantity of concrete to be removed prior to a repair is not an easy task, especially if it is intended that only unsound concrete be removed. Substantial overruns have been common. Errors in estimating the removal quantity can be minimized by a thorough condition survey as close as possible to the time the repair work is executed. When, by necessity, the condition survey is done far in advance of the repair work, the estimated quantities should be increased to account for continued deterioration.

n. Vibration and damage control. Blasting operations in or adjacent to buildings, structures, or other facilities should be carefully planned with full consideration of all forces and conditions involved. Appropriate vibration

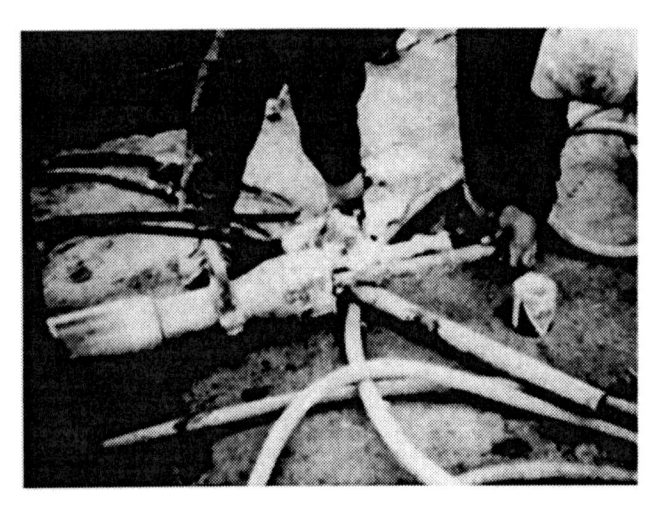

Figure 5-14. Plug-and-feather splitter

and damage control should be established in accordance with EM 385-1-1.

5-3. Preparation for Repair

One of the most important steps in the repair or rehabilitation of a concrete structure is the preparation of the surface to be repaired. The repair will only be as good as the surface preparation, regardless of the nature or sophistication (expense) of the repair material. For reinforced concrete, repairs must include proper preparation of the reinforcing steel to develop bond with the replacement concrete to ensure desired behavior in the structure. Preparation of concrete and reinforcing steel after removal of deteriorated concrete and anchor systems are discussed in the following.

a. Concrete surfaces.

(1) General considerations.

(a) The desired condition of the concrete surface immediately before beginning a repair depends somewhat on the type of repair being undertaken. For example, a project involving the application of a penetrating sealer may require only a broom-cleaned dry surface, whereas another project involving the placement of a latex-modified concrete overlay may require a sound, clean, rough-textured, wet surface. However, the desired condition of the prepared surface for most repairs will be sound, clean, rough-textured, and dry.

(b) Concrete is removed to a fixed depth for many maintenance and repair projects, leaving local areas of deteriorated concrete that must be removed as part of the surface preparation work. This secondary removal is typically accomplished with hand-held impact tools. Boom-mounted breakers and rotary-head milling are frequently used to remove nonreinforced concrete where extensive amounts of secondary removal are required.

(c) In most concrete repair projects, all damaged or deteriorated material should be removed. However, it is not always easy to determine when all such material has been removed. The best recommendation is to continue to remove material until aggregate particles are being broken rather than simply being removed from the cement matrix.

(d) Whenever concrete is removed with impact tools or by rotary-head milling, there is the potential for very small-scale damage to the surface of the concrete left in place. Unless this damaged layer is removed, the replacement material will suffer what appears to be a bond failure. Thus, a perfectly sound and acceptable replacement material may fail because of improper surface preparation.

(e) Following secondary removal, all exposed surfaces should be prepared with dry or wet sandblasting or water-jet blasting to remove any damaged surface material. Surfaces that were exposed by water-jet blasting will typically not require this surface preparation.

(2) Methods of surface preparation.

(a) Chemical cleaning. In cases in which concrete is contaminated with oil, grease, or dirt, these contaminants must be removed prior to placement of repair materials. Detergents, trisodium phosphate, and various other proprietary concrete cleaners are available for this work. It is also important that all traces of the cleaning agent be removed after the contaminating material is removed. Solvents should not be used to clean concrete since they dissolve the contaminants and carry them deeper into the concrete. Muriatic acid, commonly used to etch concrete surfaces, is relatively ineffective for removing grease or oil.

(b) Mechanical cleaning. There is a variety of mechanical devices available for cleaning concrete surfaces. These devices include scabblers, scarifiers, and impact tools. Depending upon the hammer heads used or the nature of the abrasive material, a variety of degrees of surface preparation may be achieved. After use of one of these methods, it may be necessary to use another means (waterjetting or wet sandblasting) for final cleaning of the surface.

(c) Shot blasting. Steel shot blasting produces a nearly uniform profile that is ideally suited for thin overlay repairs. It can produce light-brush blasting to 6-mm (1/4-in.)-depth removal depending on the size shot selected and the duration of the removal effort. The debris is vacuumed up and retained by the unit. Steel shot blasting leaves the surface dry for immediate application of a bonding agent, coating, or overlay.

(d) Blast cleaning. Blast cleaning includes wet and dry sandblasting, and water jetting. When sandblasting is used, the air source must be equipped with an effective oil trap to prevent contamination of the concrete surface during the cleaning operation. Water-jetting equipment with operating pressures of 40 to 70 MPa (6,000 to 10,000 psi) is commercially available for cleaning concrete. This equipment is very effective when used as the final step in surface preparation.

(e) Acid etching. Acid etching of concrete surfaces has long been used to remove laitance and normal amounts of dirt. The acid will remove enough cement paste to provide a roughened surface which will improve the bond of replacement materials. ACI 515.1R recommends that acid etching be used only when no alternative means of surface preparation can be used. The preparation methods described earlier are believed to be more effective than acid treatment. If acid is used, the surface should be cleaned of grease and oil with appropriate agents, and the cleaning agents should be rinsed off the surface before the acid is added. Acid is then added at a rate of approximately 1 L/sq m (1 qt/sq yd), and it should be worked into the concrete surface with a stiff brush or broom. When the foaming stops (3 to 5 min), the acid should be rinsed off, and brooms should be used to remove reaction products and any loosened particles. The surface should be checked with litmus or pH paper to determine that all acid has been removed.

(f) Bonding agents. The general guidance is that small thin patches (less than 50 mm (2 in.) thick) should receive a bonding coat while thicker replacements probably do not require any bonding agent. Excellent bond of fresh-to-hardened concrete can be achieved with proper surface preparation and without the use of bonding agents. The most common bonding agents are simply grout mixtures of cement slurry or equal volumes of portland cement and fine aggregate mixed with water to the consistency of thick cream. The grout must be worked into the surface with stiff brooms or brushes. The grout should not be allowed to dry out before the concrete is placed. A maximum distance of 1.5 m (5 ft) or a period of 10 min ahead of the concrete placement are typical figures used in the specification. There is a wide variety of epoxy and other polymer bonding agents available. If one of these products is used, the manufacturer's recommendations must be followed. Improperly applied bonding agents can actually reduce bond.

b. Reinforcing steel.

(1) General considerations.

(a) By far, the most frequent cause of damage to reinforcing steel is corrosion. Other possible causes of damage are fire and chemical attack. The same basic preparation and repair procedures may be used for all of these causes of damage.

(b) Once the cause and the magnitude of the damage have been determined, it remains to expose the steel, evaluate its structural condition, and prepare the reinforcement for the placement of the repair material. Proper steps to prepare the reinforcement will ensure that the repair method is a permanent solution rather than a temporary solution that will deteriorate in a short period of time.

(2) Removal of concrete surrounding reinforcing steel. The first step in preparing reinforcing steel for repair is the removal of the deteriorated concrete surrounding the steel. Usually, the deteriorated concrete above the top reinforcement can be removed with a jackhammer. For this purpose, a light (14-kg (30-lb)) hammer should be sufficient and should not significantly damage sound concrete at the periphery of the damaged area. Extreme care should be exercised to ensure that further damage to the reinforcing steel is not inflicted in the process of removing the deteriorated concrete. Jackhammers can heavily damage reinforcing steel if the hammer is used without knowledge of the location of the steel. For this reason, a copy of the structural drawings should be used to determine where the reinforcement is located and its size, and a pathometer should be used to determine the depth of the steel in the concrete. Once the larger pieces of the damaged concrete have been removed, a (7-kg (15-lb)) chipping hammer should be used to remove the concrete in the vicinity of the reinforcement. Water-jet blasting may also be used for removal of concrete surrounding the reinforcing steel.

(3) How much concrete to remove. Obviously, all weak, damaged, and easily removable concrete should be chipped away. If more than one-half of the perimeter of the bar has been exposed during removal of deteriorated concrete, then concrete removal should continue to give a clear space behind the reinforcing steel of 6 mm (1/4 in.) plus the dimension of the maximum size aggregate. If less than one-half of the perimeter of a bar is exposed after concrete removal, the bar should be inspected, cleaned as necessary, and then repairs should proceed without further concrete removal. However, if inspection indicates that a bar or bars must be replaced, concrete must be removed to give the clear space indicated above.

(4) Inspection of reinforcing steel. Once deteriorated concrete has been removed, reinforcing steel should be carefully inspected. If the cross-sectional area of a bar has been significantly reduced by corrosion or other

means, the steel may have to be replaced. If there is any question concerning the ability of the steel to perform as designed, a structural engineer should be consulted. Project specifications should include a provision whereby decisions concerning repair versus replacement of reinforcing steel can be made during the project as the steel is exposed.

(5) Replacing reinforcing steel. The easiest method of replacing reinforcement is to cut out the damaged area and splice in replacement bars. A conventional lap splice is preferred. The requirements for length of lap should conform to the requirements of ACI 318. If mechanical splices are considered, their use should be approved by a structural engineer. If a welded splice is used, it should also be performed in accordance with ACI 318. Butt welding should be avoided because of the high degree of skill required to perform a full penetration weld. High-strength steel should not be welded.

(6) Cleaning reinforcing steel.

(a) When it has been determined that the steel does not need replacing, the steel should be thoroughly cleaned of all loose rust and foreign matter before the replacement concrete is placed. For limited areas, wire brushing or other hand methods of cleaning are acceptable. For larger areas, dry sandblasting is the preferred method. The sandblasting must remove all the rust from the underside of the reinforcing bars. Normally, the underside is not directly hit by the high-pressure sand particles and must rely on rebound force as the sand comes off the substrate concrete surface. The operator must be suited with a respiratory device because of the health hazard associated with dry blasting.

(b) The type of air compressor used in conjunction with sandblasting is important. When the steel is cleaned and loose particles are blown out of the patch area after cleaning, it is important that neither the reinforcing steel nor the concrete substrate surface be contaminated with oil from the compressor. For this reason, either an oil-free compressor or one that has a good oil trap must be used.

(c) Alternative methods of cleaning the steel are wet sandblasting or water-jet blasting. These methods are not as good as dry sandblasting, because they provide the water and oxygen necessary to begin the corrosion process again once the steel has been cleaned.

(d) There is always the possibility that freshly cleaned reinforcing steel will rust between the time it is cleaned and the time that the next concrete is placed. If the rust that forms is tightly bonded to the steel, there is no need to take further action. If the rust is loosely bonded or in any other way may inhibit bonding between the steel and the concrete, the reinforcing bars must be cleaned again immediately before concrete placement.

c. Anchors. Dowels may be required in some situations to anchor the repair material to the existing concrete substrate. ACI 355.1R summarizes anchor types and provides an overview of anchor performance and failure modes under various loading conditions. It also covers design and construction considerations and summarizes existing requirements in codes and specifications. Design criteria for anchoring relatively thin sections (less than 0.8 m (2.5 ft)) of cast-in-place concrete are described in Section 8-1. Anchor installation underwater is discussed in Section 8-6. Most of the anchors used in repair are installed in holes drilled in the concrete substrate and can be classified as either bonded or expansion anchors.

(1) Drilling. Anchor holes should be drilled with rotary carbide-tipped or diamond-studded bits or hand-hammered star drill bits. Drilling with a jackhammer is not recommended because of the damage that results immediately around the hole from the impact. Holes should be cleaned with compressed air and plugged with a rag or other suitable material until time for anchor installation. Holes should be inspected for proper location, diameter, depth, and cleanliness prior to installation of anchors.

(2) Bonded anchors. Bonded anchors are headed or headless bolts, threaded rods, or deformed reinforcing bars. Bonded anchors are classified as either grouted anchors or chemical anchors.

(a) Grouted anchors are embedded in predrilled holes with neat portland cement, portland cement and sand, or other commercially available premixed grout. An expansive grout additive and accelerator are commonly used with cementitious grouts.

(b) Chemical anchors are embedded in predrilled holes with two-component polyesters, vinylesters, or epoxies. The chemicals are available in four forms: glass capsules, plastic cartridges, tubes ("sausages"), or bulk. Following insertion into the hole, the glass capsules and tubes are both broken and their contents mixed by insertion and spinning of the anchor. The plastic cartridges are used with a dispenser and a static mixing nozzle to mix the two components as they are placed in the drill hole. Bulk systems are predominately epoxies which are mixed in a pot, or pumped through a mixer and injected into the hole after which the anchor is immediately inserted.

(c) Some chemical grouts creep under sustained loading, and some lose their strength when exposed to temperatures over 50 °C (120 °F). Creep tests were conducted, as part of the REMR Research Program, by subjecting anchors to sustained loads of 60 percent of their yield strength for 6 months. The slippage exhibited by anchors embedded in polyester resin was approximately 30 times higher than that of anchors embedded in portland-cement (Best and McDonald 1990b).

(3) Expansion anchors. Expansion anchors are designed to be inserted into predrilled holes and then expanded by either tightening a nut, hammering the anchor, or expanding into an undercut in the concrete. Expansion anchors that rely on side point contact to create frictional resistance should not be used where anchors are subjected to vibratory loads. Some wedge-type anchors perform poorly when subjected to impact loads. Undercut anchors are suitable for dynamic and impact loads.

(4) Load tests. Following installation, randomly selected anchors should be tested to ensure compliance with the specifications. In some field tests, anchors have exhibited significant slippage prior to achieving the desired tensile capacity. Therefore, it may be desirable to specify a maximum displacement in addition to the minimum load capacity.

Chapter 6
Materials and Methods for Repair and Rehabilitation

6-1. Introduction

This chapter contains descriptions of various materials and methods that are available for repair or rehabilitation of concrete structures. Each of the entries in this chapter will include description, applications and limitations, and procedure. Although the repair procedures given in this chapter are current practice, they may not be used directly in project specifications because each repair project may require unique remedial action. Emmons (1993) provides a discussion of materials and methods for concrete repair with extensive, detailed illustrations.

6-2. Additional Reinforcement

a. Description. Additional reinforcement, as the name implies, is the provision of additional reinforcing steel, either conventional reinforcement or prestressing steel, to repair a cracked concrete section. In either case, the steel that is added is to carry the tensile forces that have caused cracking in the concrete.

b. Applications and limitations. Cracked reinforced concrete bridge girders have been successfully repaired by use of additional conventional reinforcement (Stratton, Alexander, and Nolting 1982). Posttensioning is often the desirable solution when a major portion of a member must be strengthened or when the cracks that have formed must be closed. For the posttensioning method, some form of abutment is needed for anchorage, such as a strongback bolted to the face of the concrete, or the tendons can be passed through and anchored in connecting framing.

c. Procedure.

(1) Conventional reinforcement.

(a) This technique consists of sealing the crack, drilling holes 19 mm (3/4 in.) in diam at 90 deg to the crack plane (Figure 6-1), cleaning the hole of dust, filling the hole and crack plane with an adhesive (typically epoxy) pumped under low pressure 344 to 552 KPa (50 to 80 psi), and placing a reinforcing bar into the drilled hole. Typically, No. 4 or 5 bars are used, extending at least 0.5 m (1.6 ft) on each side of the crack. The adhesive

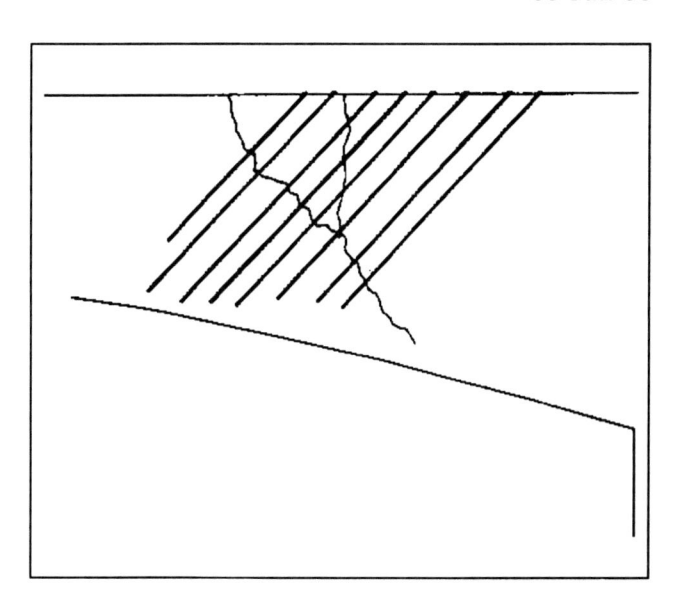

Figure 6-1. Crack repair using conventional reinforcement with drillholes 90 deg to the crack plane

bonds the bar to the walls of the hole, fills the crack plane, bonds the cracked concrete surfaces together in one monolithic form, and thus reinforces the section.

(b) A temporary elastic crack sealant is required for a successful repair. Gel-type epoxy crack sealants work very well within their elastic limits. Silicone or elastomeric sealants work well and are especially attractive in cold weather or when time is limited. The sealant should be applied in a uniform layer approximately 1.6 to 2.4 mm (1/16 to 3/32 in.) thick and should span the crack by at least 19 mm (3/4 in.) on each side.

(c) Epoxy adhesives used to rebond the crack should conform to ASTM C 881, Type I, low-viscosity grade.

(d) The reinforcing bars can be spaced to suit the needs of the repair. They can be placed in any desired pattern, depending on the design criteria and the location of the in-place reinforcement.

(e) Concrete elements may also be reinforced externally by placement of longitudinal reinforcing bars and stirrups or ties around the members and then encasing the reinforcement with shotcrete or cast-in-place concrete. Also, girders and slabs have been reinforced by addition of external tendons, rods, or bolts which are prestressed. The exterior posttensioning is performed with the same equipment and design criteria of any posttensioning project. If desirable for durability or for esthetics, the exposed posttensioning strands may be covered by concrete.

(2) Prestressing steel. This technique uses prestressing strands or bars to apply a compressive force (Figure 6-2). Adequate anchorage must be provided for the prestressing steel, and care is needed so that the problem will not merely migrate to another part of the structure. The effects of the tensioning force (including eccentricity) on the stress within the structure should be carefully analyzed. For indeterminate structures posttensioned according to this procedure, the effects of secondary moments and induced reactions should be considered.

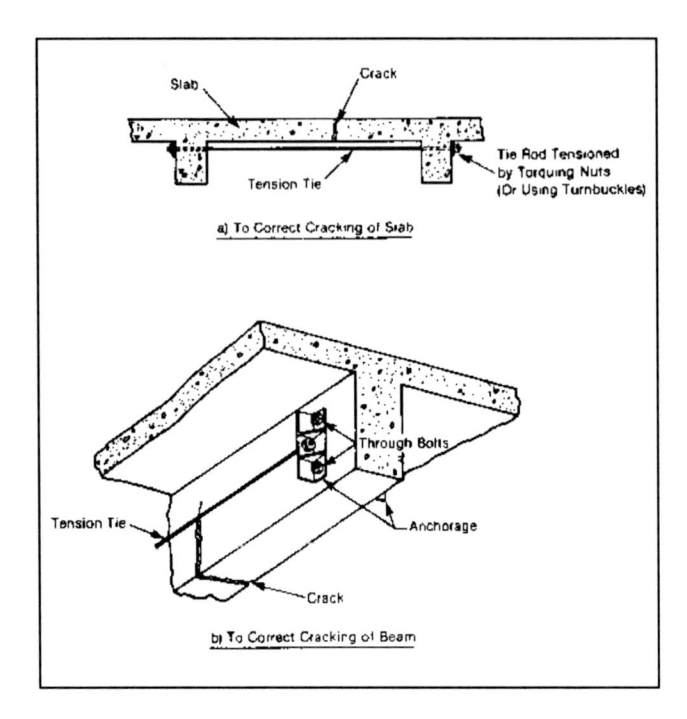

Figure 6-2. Crack repair with use of external prestressing strands or bars to apply a compressive force

(3) Steel plates. Cracks in slabs on grade have been repaired by making saw cuts 50 to 75 mm (2 to 3 in.) deep across the crack and extending 150 to 300 mm (6 to 12 in.) on either side of the crack, filling the saw cuts and the crack with epoxy, and forcing a steel plate of appropriate size into each saw cut.

6-3. Autogenous Healing

a. Description. Autogenous healing is a natural process of crack repair that can occur in the presence of moisture and the absence of tensile stress (Lauer 1956).

b. Applications and limitations. Autogenous healing has practical application for closing dormant cracks in a moist environment. Healing will not occur if the crack is active and is subjected to movement during the healing period. Healing will also not occur if there is a positive flow of water through the crack which dissolves and washes away the lime deposit. A partial exception is a situation in which the flow of water is so slow that complete evaporation occurs at the exposed face causing redeposition of the dissolved salts.

c. Mechanism. Healing occurs through the carbonation of calcium hydroxide in the cement paste by carbon dioxide, which is present in the surrounding air and water. Calcium carbonate and calcium hydroxide crystals precipitate, accumulate, and grow within the cracks. The crystals interlace and twine, producing a mechanical bonding effect, which is supplemented by chemical bonding between adjacent crystals and between the crystals and the surfaces of the paste and the aggregate. As a result, some of the tensile strength of the concrete is restored across the cracked section, and the crack may become sealed. Saturation of the crack and the adjacent concrete with water during the healing process is essential for developing any substantial strength. Continuous saturation accelerates the healing. A single cycle of drying and reimmersion will produce a drastic reduction in the amount of healing.

6-4. Conventional Concrete Placement

a. Description. This method consists of replacing defective concrete with a new conventional concrete mixture of suitable proportions that will become an integral part of the base concrete. The concrete mixture proportions must provide for good workability, strength, and durability. The repair concrete should have a low w/c and a high percentage of coarse aggregate to minimize shrinkage cracking.

b. Applications and limitations. If the defects in the structure go entirely through a wall or if the defects go beyond the reinforcement and if the defective area is large, then concrete replacement is the desired method. Replacement is sometimes necessary to repair large areas of honeycomb in new construction. Conventional concrete should not be used for replacement in areas where an aggressive factor which has caused the deterioration of the concrete being replaced still exists. For example, if the deterioration noted has been caused by acid attack, aggressive-water attack, or even abrasion-erosion, it is doubtful that repair by conventional-concrete placement will be successful unless the cause of deterioration is removed. Concrete replacement methods for repairing lock walls and stilling basins are given in Sections 8-1 and 8-3, respectively, and repair by placing a thin concrete overlay is discussed in Section 6-17.

c. Procedure.

(1) Concrete removal is always required for this type of repair. Removal of affected areas should continue until there is no question that sound concrete has been reached. Additional chipping may be necessary to attain a satisfactory depth (normally 150 mm (6 in.) or more) and to shape the cavity properly. Final chipping should be done with a light hammer to remove any unsound concrete that remains. In a vertical surface (Figure 6-3), the cavity should have the following:

(a) A minimum of spalling or featheredging at the periphery of the repair area.

(b) Vertical sides and horizontal top at the surface of the member (the top line of the cavity may be stepped).

(c) Inside faces generally normal to the formed surface, except that the top should slope up toward the front at about a 1:3 slope.

(d) Keying as necessary to lock the repair into the structure.

(e) Sufficient depth to reach at least 6 mm (1/4 in.) plus the dimension of the maximum size aggregate behind any reinforcement.

(f) All interior corners rounded with a radius of about 25 mm (1 in.).

(2) Surfaces must be thoroughly cleaned by sandblasting (wet or dry), shotblasting, or another equally satisfactory method, followed by final cleaning with compressed air or water. Sandblasting effects should be confined to the surface that is to receive the new concrete. Dowels and reinforcement are often installed to make the patch self-sustaining and to anchor it to the underlying concrete, thus providing an additional safety factor.

(3) Forming will usually be required for massive repairs in vertical surfaces. The front form and the back form, where one is required, should be substantially constructed and mortartight. The back form may be assembled in one piece, but the front panel should be constructed as placing progresses so that the concrete can be conveniently placed in lifts. The contact surface should be dry at the time of patching. Small, thin repairs (less than 50 mm (2 in.) thick) should receive a bonding coat while thicker placements usually do not require a bonding coat (see paragraph 5-3*a*(2)(f)). The surface is first carefully coated with a thin layer of mortar, not exceeding 3 mm (1/8 in.) in thickness, containing sand passing the No. 16 sieve, and having the same w/c as the concrete to be used in the replacement. Hand-rubbing the mortar into the surface is effective. Epoxy resin

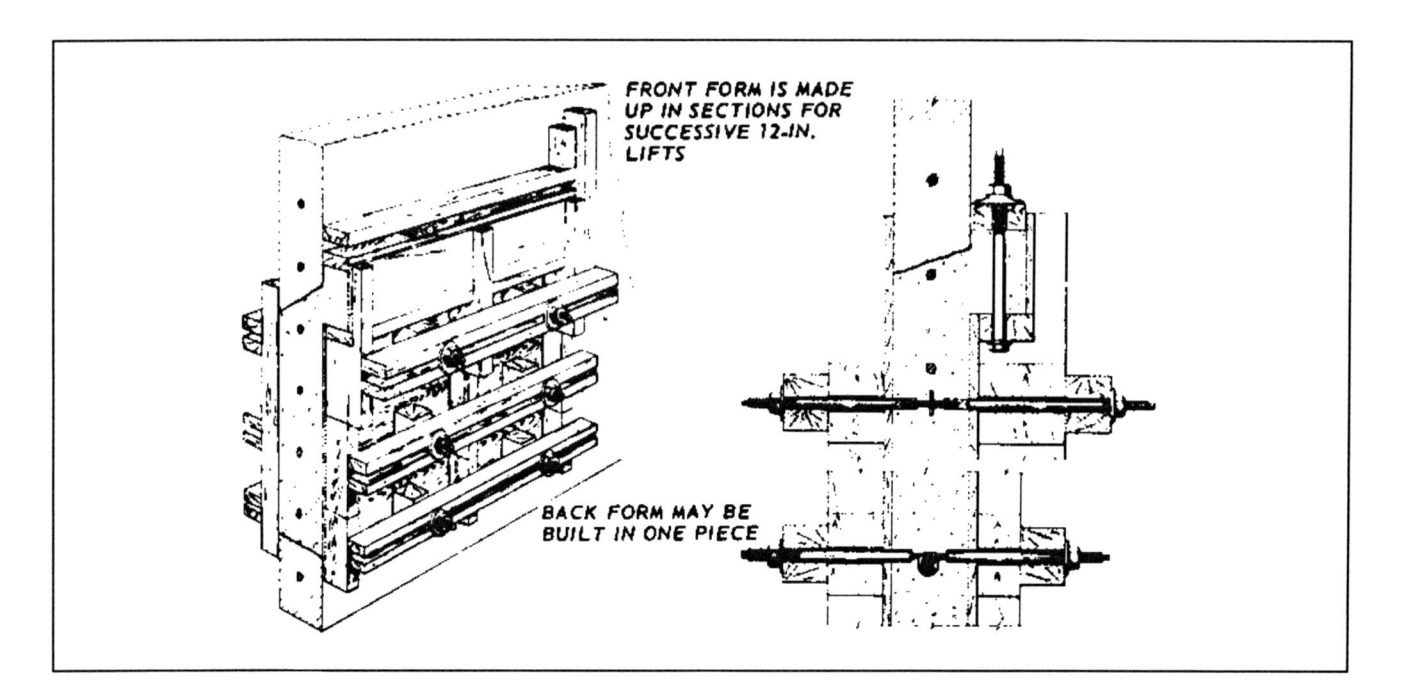

Figure 6-3. Detail of form for concrete replacement in walls after removal of all unsound concrete

meeting ASTM C 881, Type II or Type V may also be used. ACI 503.2 provides a standard specification for bonding plastic concrete to hardened concrete with epoxy adhesives.

(4) Concrete used for repair should conform to EM 1110-2-2000. To minimize strains caused by temperature, moisture change, shrinkage, etc., concrete for the repair should generally be similar to the old concrete in maximum size of aggregate and w/c. Each lift should be thoroughly vibrated. Internal vibration should be used except where accessibility and size of placement will not allow it. If internal vibration can not be used, external vibration may be used. If external vibration must be used, placement through a chimney, followed by a pressure cap (Figure 6-3) should be required. If good internal vibration can be accomplished, the pressure cap may not be needed. The slump should be as low as practical, and a chimney and pressure cap should be used. A tighter patch results if the concrete is placed through a chimney at the top of the front form.

(5) When external vibration is necessary, immediately after the cavity has been filled, a pressure cap should be placed inside the chimney (Figure 6-3). Pressure should be applied while the form is vibrated. This operation should be repeated at 30-min intervals until the concrete hardens and no longer responds to vibration. The projection left by the chimney should normally be removed the second day. Proper curing is essential.

(6) The form and pump technique is often used to place conventional concrete (or other materials) in vertical or over head applications. The proper size variable output concrete pump is used to pump concrete into a cavity confined by formwork. Care must be taken to trim the original concrete surfaces that may entrap air, or these areas may be vented. Forming must be nearly watertight and well braced so that pressure from the pumps can help achieve bonding of the new concrete to the old.

(7) Curing of concrete repairs is very important, especially if relatively thin repairs are made in hot weather. Shrinkage cracks can develop quickly under such conditions. Moist curing conforming to the guidelines in EM 1110-2-2000 is the preferred curing method.

6-5. Crack Arrest Techniques

a. Description. Crack arrest techniques are those procedures that may be used during the construction of a massive concrete structure to stop crack propagation into subsequent concrete lifts.

b. Applications and limitations. These techniques should be used only for cracking caused by restrained volume change of the concrete. They should not be used for cracking caused by excessive loading.

c. Procedure. During construction of massive concrete structures, contraction cracks may develop as the concreting progresses. Such cracks may be arrested by use of the following techniques.

(1) The simplest technique is to place a grid of reinforcing steel over the cracked area. The reinforcing steel should be surrounded by conventional concrete rather than the mass concrete being used in the structure.

(2) A somewhat more complex procedure is to use a piece of semicircular pipe as shown in Figure 6-4. The installation procedure is as follows: First, the semicircular pipe is made by splitting a 200-mm (8-in.)-diam piece of 16-gauge pipe and bending it to a semicircular shape with about a 76-mm- (3-in.-) flange on each side. Then, the area surrounding the crack should be well cleaned and the pipe should be centered on the crack. Once in place, the sections of the pipe should be welded together. Holes should be cut into the pipe to receive grout pipes. Finally, the pipe section should be covered with concrete placed concentrically by hand methods. The grout pipes may be used for grouting at a later date to attempt to restore structural integrity of the cracked section.

(3) A piece of bond-breaking membrane placed on a construction joint over the crack has been used with varying degrees of success.

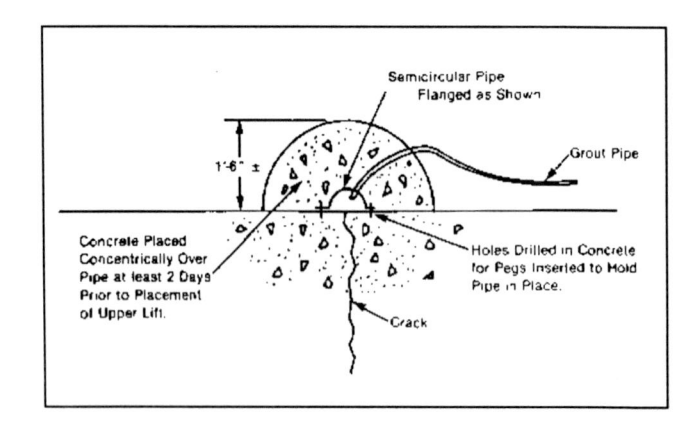

Figure 6-4. The use of a semicircular pipe in the crack arrest method of concrete repair

6-6. Drilling and Plugging

a. Description. Drilling and plugging a crack consists of drilling down the length of the crack and grouting it to form a key (Figure 6-5).

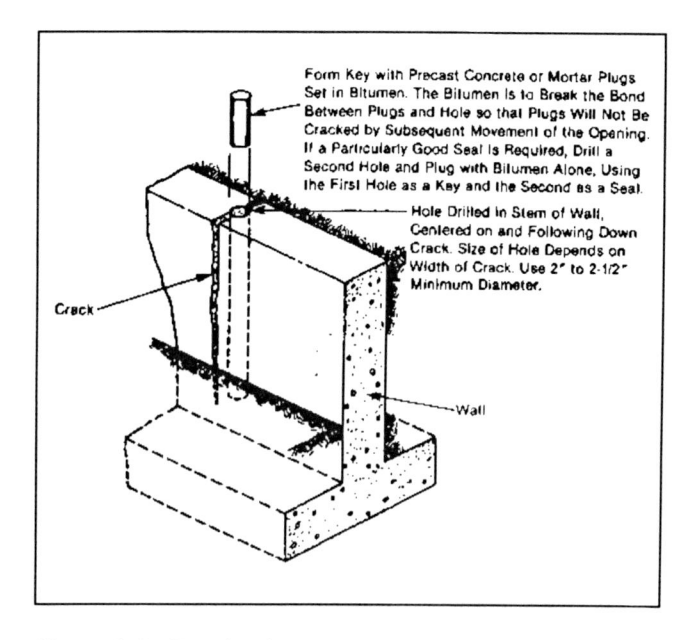

Form Key with Precast Concrete or Mortar Plugs Set in Bitumen. The Bitumen is to Break the Bond Between Plugs and Hole so that Plugs Will Not Be Cracked by Subsequent Movement of the Opening. If a Particularly Good Seal is Required, Drill a Second Hole and Plug with Bitumen Alone, Using the First Hole as a Key and the Second as a Seal.

Hole Drilled in Stem of Wall, Centered on and Following Down Crack. Size of Hole Depends on Width of Crack. Use 2″ to 2-1/2″ Minimum Diameter.

Crack

Wall

Figure 6-5. Repair of crack by drilling and plugging

b. Applications and limitations. This technique is applicable only where cracks run in reasonably straight lines and are accessible at one end. This method is most often used to repair vertical cracks in walls.

(1) Procedure. A hole (typically 50 to 75 mm (2 to 3 in.) in diam) should be drilled, centered on, and following the crack. The hole must be large enough to intersect the crack along its full length and provide enough repair material to structurally take the loads exerted on the key. The drilled hole should then be cleaned and filled with grout. The grout key prevents transverse movement of the sections of concrete adjacent to the crack. The key will also reduce heavy leakage through the crack and loss of soil from behind a leaking wall.

(2) If watertightness is essential and structural load transfer is not, the drilled hole should be filled with a resilient material of low modulus such as asphalt or polyurethane foam in lieu of portland-cement grout. If the keying effect is essential, the resilient material can be placed in a second hole, the first being grouted.

6-7. Drypacking

a. Description. Drypacking is a process of ramming or tamping into a confined area a low water-content mortar. Because of the low w/c material, there is little shrinkage, and the patch remains tight and is of good quality with respect to durability, strength, and watertightness. This technique has an advantage in that no special equipment is required. However, the method does require that the craftsman making the repair be skilled in this particular type of work.

b. Applications and limitations. Drypacking can be used for patching rock pockets, form tie holes, and small holes with a relatively high ratio of depth to area. It should not be used for patching shallow depressions where lateral restraint cannot be obtained, for patching areas requiring filling in back of exposed reinforcement, nor for patching holes extending entirely through concrete sections. Drypacking can also be used for filling narrow slots cut for the repair of dormant cracks. The use of drypack is not recommended for filling or repairing active cracks.

c. Procedure.

(1) The area to be repaired should be undercut slightly so that the base width is slightly greater than the surface width. For repairing dormant cracks, the portion adjacent to the surface should be widened to a slot about 25 mm (1 in.) wide and 25 mm (1 in.) deep. This is most conveniently done with a power-driven sawtooth bit. The slot should also be undercut slightly. After the area or slot is thoroughly cleaned and dried, a bond coat should be applied. Placing of the drypack mortar should begin immediately. The mortar usually consists of one part cement, two and one-half to three parts sand passing a No. 16 sieve, and only enough water so that the mortar will stick together when molded into a ball by slight pressure of the hands and will not exude water but will leave the hands dry. Latex-modified mortar is being increasingly used in lieu of straight portland-cement mortar. Preshrunk mortar may be used to repair areas too small for the tamping procedure. Preshrunk mortar is a low water-content mortar that has been mixed and allowed to stand idle 30 to 90 min, depending on the temperature, prior to use. Remixing is required after the idle period.

(2) Drypack mortar should be placed in layers having a compacted thickness of about 10 mm (3/8 in.).

Each layer should be compacted over its entire surface by use of a hardwood stick. For small areas, the end of the stick is placed against the mortar and tamping is begun at the middle of the area and progresses toward the edges to produce a wedging effect. For larger areas, a T-shaped rammer may be used; the flat head of the T is placed against the mortar and hammered on the stem. It is usually necessary to scratch the surface of the compacted layers to provide bond for the next layer. Successive layers of drypack are placed without interval, unless the material becomes spongy, in which case there should be a short wait until the surface stiffens. Areas should be filled flush and finished by striking a flat-sided board or the flat of the hardwood stick against the surface. Steel trowelling is not suitable. After being finished, the repaired area should be cured. If the patch must match the color of the surrounding concrete, a blend of portland cement and white cement may be used. Normally, about one-third white cement is adequate, but the precise proportions can only be determined by trial.

6-8. Fiber-Reinforced Concrete

   a. Description. Fiber-reinforced concrete is composed of conventional portland-cement concrete containing discontinuous discrete fibers. The fibers are added to the concrete in the mixer. Fibers are made from steel, plastic, glass, and other natural materials. A convenient numerical parameter describing a fiber is its aspect ratio, defined as the fiber length divided by an equivalent fiber diameter. Typical aspect ratios range from about 30 to 150 for lengths of 6.4 to 76 mm (0.25 to 3 in.).

   b. Applications and limitations. Fiber-reinforced concrete has been used extensively for pavement repair. Fiber-reinforced concrete has been used to repair erosion of hydraulic structures caused by cavitation or high velocity flow and impact of large debris (ACI 210R). However, laboratory tests and field experience show that the abrasion-erosion resistance of fiber-reinforced concrete is significantly less than that of conventional concrete with the same w/c and aggregate type (Liu 1980, Liu and McDonald 1981). The slump of a concrete mixture is significantly reduced by the addition of fibers. Use of the inverted slump cone test for workability is recommended. Reliance on slump tests often results in the use of excessive water in an attempt to maintain a slump, without improving workability. A fiber mixture will generally require more vibration to consolidate the concrete.

   c. Procedure. Preparation of the area to be repaired, mixing, transporting, placing, and finishing fiber-reinforced concrete follows the procedures for and generally uses the same equipment as plain concrete (ACI 544.3R). Pumping of steel fiber-reinforced concrete with up to 1.5 percent fibers by volume has been done successfully. Three-pronged garden forks are preferable to shovels for handling the fiber-reinforced concrete. Mixture design and especially the amount of fibers used are critical so that design parameters for strength and durability are met and the mixture will still be workable. About 2 percent by volume is considered a practical upper limit for field placement with the necessary workability. Steel fiber-reinforced shotcrete, with up to 2.0 percent fibers by volume, generally mixed with the dry-mixture process has been successfully used to repair concrete. Polypropylene fibers have been added to acrylic polymer modified concrete for repair of a lockwall (Dahlquist 1987).

6-9. Flexible Sealing

   a. Description. Flexible sealing involves routing and cleaning the crack and filling it with a suitable field-molded flexible sealant. This technique differs from routing and sealing in that, in this case, an actual joint is constructed, rather than a crack simply being filled.

   b. Applications and limitations. Flexible sealing may be used to repair major, active cracks. It has been successfully used in situations in which there is a limited water head on the crack. This repair technique does not increase the structural capacity of the cracked section. Another process used to form a flexible joint from an active or inactive water-filled crack is described in Section 6-11. This process may be used in lieu of or in addition to flexible sealing. Chemical grouting is a more complicated and expensive procedure, but it can be used in conditions of flowing water.

   c. Procedure. Active cracks can be routed out; cleaned by sandblast or air-water jet, or both; and filled with a suitable field-molded flexible sealant (ACI 224.1R). As nearly as is practical, the sealant reservoir (slot) formed by routing should comply with the requirements for width and shape factor of a joint having equivalent movement. The selection of a suitable sealant and installation method should follow that for equivalent joints (ACI 504R).

   (1) A bond breaker should be provided at the bottom of the slot to allow the sealant to change shape without a concentration of stress on the bottom (Figure 6-6). The bond breaker may be a polyethylene strip, pressure sensitive tape, or other material which will not bond to the sealant before or during cure.

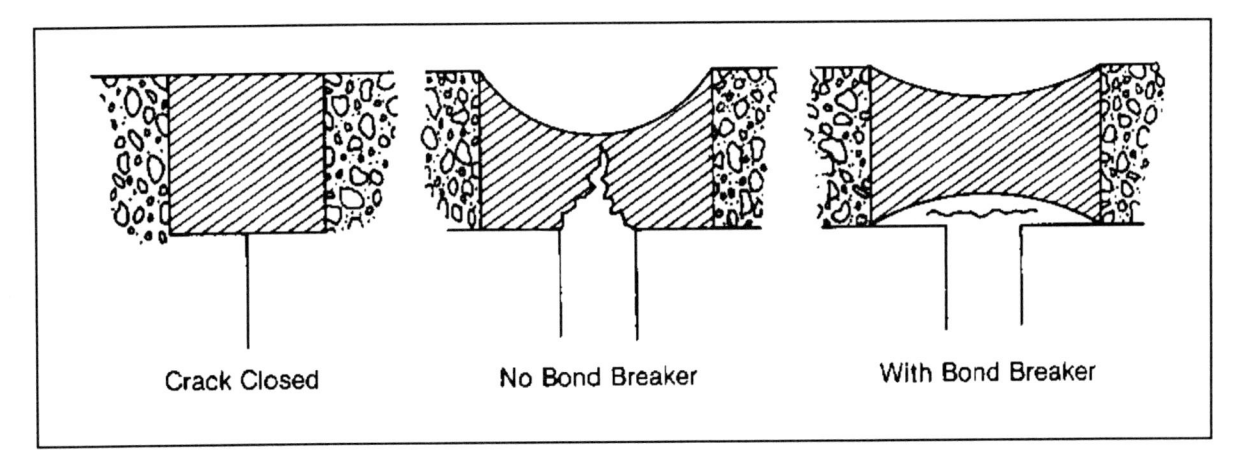

Figure 6-6. Effect of bond breaker involving a field-molded flexible sealant

(2) If a bond breaker is used over the crack, a flexible joint sealant may be trowelled over the bond breaker to provide an adequate bonding area. This is a very economical procedure and may be used on the interior of a tank, on roofs, or other areas not subject to traffic or mechanical abuse.

(3) Narrow cracks subject to movement, where esthetics are not important, may be sealed with a flexible surface seal (Figure 6-7).

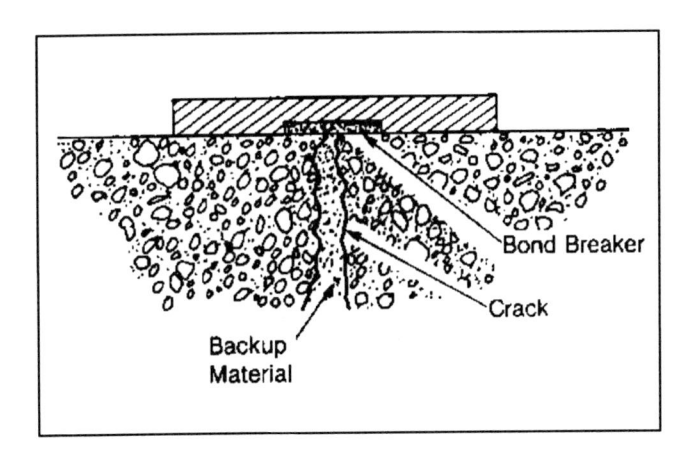

Figure 6-7. Repair of a narrow crack with flexible surface seal

(4) When repairing cracks in canal and reservoir linings or low-head hydraulic structures where water movement or pressure exists, a retaining cap must be used to confine the sealant. A simple retainer can be made by positioning a metal strip across the crack and fastening it to expandable anchors or grouted bolts installed in the concrete along one side of the crack. To maintain hydraulic efficiency in some structures, it may be necessary to cut the concrete surface adjacent to the crack and to place the retaining cap flush with the original flow lines (Figure 6-8).

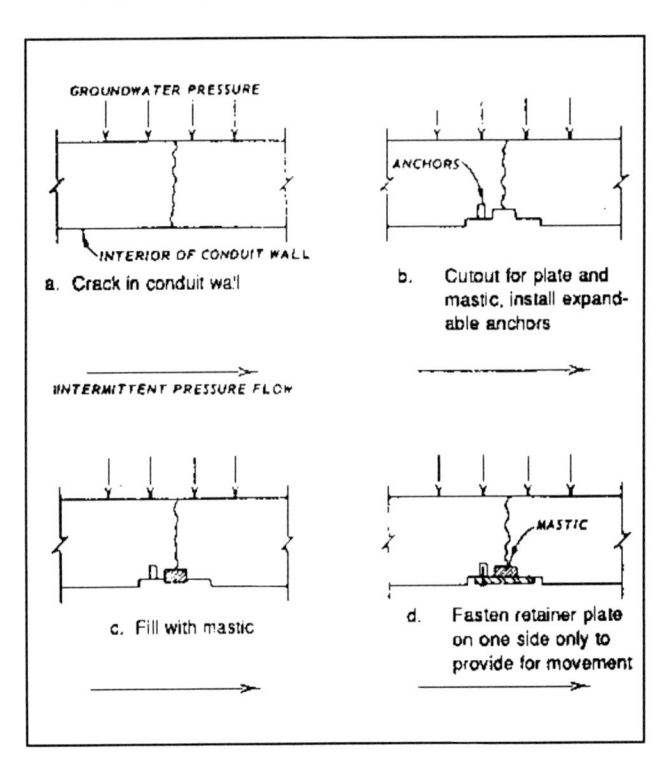

Figure 6-8. Repair of crack by use of a retainer plate to hold mastic in place against external pressure

6-10. Gravity Soak

a. Description. High molecular weight methacrylate (HMWM) is poured or sprayed onto any horizontal concrete surface and spread by broom or squeegee. The material penetrates very small cracks by gravity and capillary action, polymerizing to form a "plug" which closes off access to the reinforcing steel (Montani 1993).

   b.   *Applications and limitations.*   Repairing cracks with the gravity soak method and HMWM has become a proven and cost-effective method. Gravity soak can be an effective repair method for horizontal concrete surfaces that contain excessive, closely spaced shrinkage cracking. This would include bridge decks, parking decks, industrial floors, pavements etc. HMWM's should not be confused with methyl methacrylates (MMA's). While MMA's are very volatile and have a low flash point, HMWM's have a high flashpoint, and are quite safe to use.

   c.   *Procedure.*   New concrete must have cured for at least 1 week and must be air-dry. Air-drying is necessary after a rainfall. New concrete surfaces may simply be swept clean before application, but older surfaces will require cleaning of all oil, grease, tar, or other contaminants and sand blasting. The monomer is mixed with the catalyst and quickly poured onto the concrete surface. Two-component systems should be specified. Three-component systems are not recommended because improper mixing sequences can be dangerous. The material is spread by a broom or squeegee. Larger individual cracks can sometimes be treated by use of a squeegee bottle, in addition to the flooding. It is important that the material not be allowed to puddle so that smooth slick surfaces are formed. Tined or grooved surfaces may require use of a large napped roller to remove excess HVWM. After about 30 min of penetration time, areas of greater permeability or extensive cracking may require additional treatment. A light broadcast of sand is usually recommended after the HMWM initial penetration. Some sand will not adhere and should be removed, but the skid resistance will have been accomplished. The surface will be ready to accept traffic in 3 to 24 hr, according to the formulation used.

6-11. Grouting (Chemical)

   a.   *Description.*   Chemical grouts consist of solutions of two or more chemicals that react to form a gel or solid precipitate as opposed to cement grouts that consist of suspensions of solid particles in a fluid (EM 1110-1-3500). The reaction in the solution may be either chemical or physicochemical and may involve only the constituents of the solution or may include the interaction of the constituents of the solution with other substances encountered in the use of the grout. The reaction causes a decrease in fluidity and a tendency to solidify and fill voids in the material into which the grout has been injected.

   b.   *Applications and limitations.*   Cracks in concrete as narrow as 0.05 mm (0.002 in.) have been filled with chemical grout. The advantages of chemical grouts include their applicability in moist environments, wide limits of control of gel time, and their application in very fine fractures. Disadvantages are the high degree of skill needed for satisfactory use, their lack of strength, and, for some grouts, the requirement that the grout not dry out in service. Also some grouts are highly inflammable and cannot be used in enclosed spaces.

   c.   *Procedure.*   Guidance and information regarding the use of chemical grouts can be found in EM 1110-1-3500.

6-12. Grouting (Hydraulic-Cement)

   a.   *Description.*   Hydraulic-cement grouting is simply the use of a grout that depends upon the hydration of portland cement, portland cement plus slag, or pozzolans such as fly ash for strength gain. These grouts may be sanded or unsanded (neat) as required by the particular application. Various chemical admixtures are typically included in the grout. Latex additives are sometimes used to improve bond.

   b.   *Applications and limitations.*   Hydraulic-cement grouts may be used to seal dormant cracks, to bond subsequent lifts of concrete that are being used as a repair material, or to fill voids around and under concrete structures. Hydraulic-cement grouts are generally less expensive than chemical grouts and are better suited for large volume applications. Hydraulic cement grout has a tendency to separate under pressure and thus prevent 100 percent filling of the crack. Normally the crack width at the point of introduction should be at least 3 mm (1/8 in.). Also, if the crack cannot be sealed or otherwise confined on all sides, the repair may be only partially effective. Hydraulic-cement grouts are also used extensively for foundation sealing and treatments during new construction, but such applications are beyond the scope of this manual. See EM 1110-2-3506 for information relative to the use in these areas.

   c.   *Procedure.*   The procedure consists of cleaning the concrete along the crack, installing built-up seats (grout nipples) at intervals astride the crack to provide a pressure-tight contact with the injection apparatus, sealing the crack between the seats, flushing the crack to clean it and test the seal, and then grouting the entire area. Grout mixtures may vary in volumetric proportion from one part cement and five parts water to one part cement and one part water, depending on the width of the crack. The water-cement ratio should be kept as low as practical to maximize strength and minimize shrinkage. For small volumes, a manual injection gun may be used; for larger volumes, a pump should be used. After the crack is filled, the pressure should be maintained for several minutes to ensure good penetration.

6-13. High-Strength Concrete

a. Description. High-strength concrete is defined as concrete with a 28-day design compressive strength over 41 MPa (6,000 psi) (ACI 116R). This method is similar to an extension of the conventional concrete placement method described in Section 6-4. Chemical admixtures such as water-reducing admixtures (WRA's) and HRWRA's are usually required to achieve lower w/c and subsequently higher compressive strengths. Mineral admixtures are also frequently used. The special procedures and materials involved with producing high-strength concrete with silica fume are discussed in paragraph 6-30. Guidance on proportioning high-strength concrete mixtures is given in EM 1110-2-2000 and ACI 363R.

b. Applications and limitations. High-strength concrete for concrete repair is used to provide a concrete with improved resistance to chemical attack, better abrasion resistance, improved resistance to freezing and thawing, and reduced permeability. The material is slightly more expensive and requires greater control than conventional concrete. A special laboratory mixture design should always be required for high-strength concrete instead of a producers' standard mixture that requires field adjustments.

c. Procedure. Generally, concrete production and repair procedures are done in the same way as a conventional concrete. Selection of materials to be used should be based on the intended use of the material and the performance requirements. Curing is more critical with high-strength concrete than with normal-strength concrete. Water curing should be used, if practicable.

6-14. Jacketing

a. Description. Jacketing consists of restoring or increasing the section of an existing member (principally a compression member) by encasing it in new concrete

(Johnson 1965). The original member need not be concrete; steel and timber sections can be jacketed.

b. Applications and limitations. The most frequent use of jacketing is in the repair of piling that has been damaged by impact or is disintegrating because of environmental conditions. It is especially useful where all or a portion of the section to be repaired is underwater. When properly applied, jacketing will strengthen the repaired member as well as provide some degree of protection against further deterioration. However, if a concrete pile is deteriorating because of exposure to acidic water, for example, jacketing with conventional portland-cement concrete will not ensure against future disintegration.

c. Procedure. The removal of the existing damaged concrete or other material is usually necessary to ensure that the repair material bonds well to the original material that is left in place. If a significant amount of removal is necessary, temporary support may have to be provided to the structure during the jacketing process. Any suitable form material may be used. A variety of proprietary form systems are available specifically for jacketing. These systems employ fabric, steel, or fiberglass forms. Use of a preformed fiberglass jacket for repair of a concrete pile is shown in Figure 6-9. A steel reinforcement cage may be constructed around the damaged section. Once the form is in place, it may be filled with any suitable material. Choice of the filling material should be based upon the environment in which it will serve as well as a knowledge of what caused the original material to fail. Filling may be accomplished by pumping, by tremie placement, by preplaced aggregate techniques, or by conventional concrete placement if the site can be dewatered.

6-15. Judicious Neglect

a. Description. As the name implies, judicious neglect is the repair method of taking no action. This method does not suggest ignoring situations in which damage to concrete is detected. Instead, after a careful (i.e., "judicious") review of the circumstances the most appropriate action may be to take no action at all.

b. Applications and limitations. Judicious neglect would be suitable for those cases of deterioration in which the damage to the concrete is causing no current operational problems for the structure and which will not contribute to future deterioration of the concrete. Dormant cracks, such as those caused by shrinkage or some other

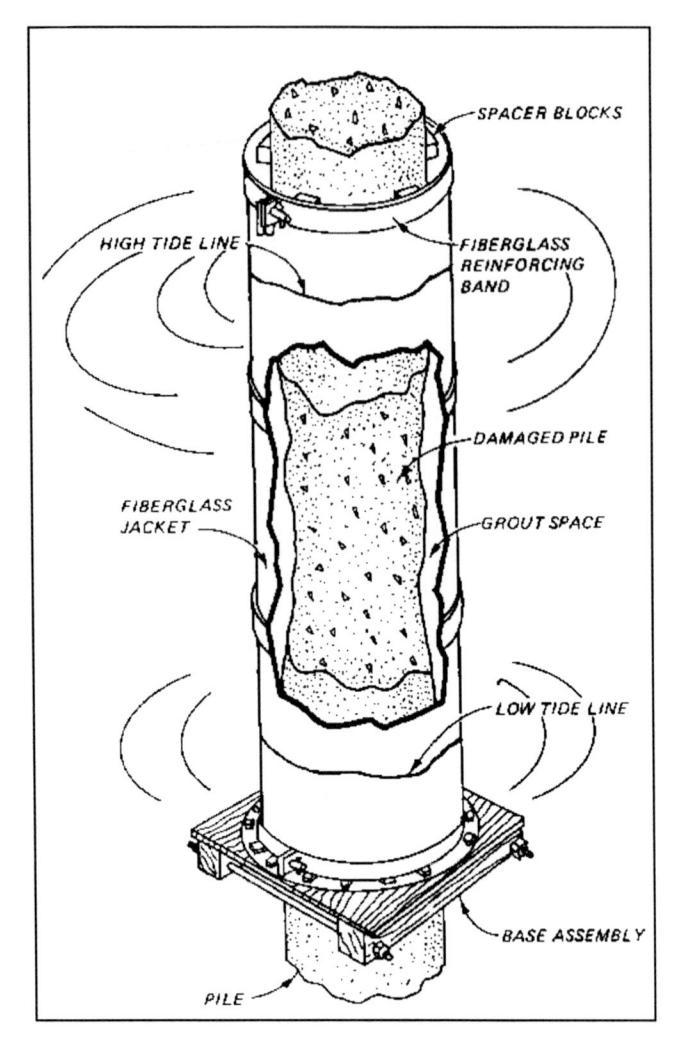

SPACER BLOCKS

HIGH TIDE LINE

FIBERGLASS
REINFORCING
BAND

DAMAGED PILE

FIBERGLASS
JACKET

GROUT SPACE

LOW TIDE LINE

BASE ASSEMBLY

PILE

Figure 6-9. Typical preformed fiberglass jacket being used in repair of a concrete pile

one-time occurrence, may be self-sealing. This does not imply an autogenous healing and gain of strength, but merely that the cracks clog with dirt, grease, or oil, or perhaps a little recrystallization occurs. The result is that the cracks are plugged, and problems which may have been encountered with leakage, particularly if leakage is the result of some intermittent cause rather than a continuing pressure head, will disappear without doing any repair.

6-16. Overlays (Polymer)

a. Description. Polymer overlays generally consist of latex-modified concrete, epoxy-modified concrete and epoxy mortar and concrete. Epoxy mortar and concrete contain aggregate and an epoxy resin binder. Latex modified concrete and epoxy modified concrete are normal portland-cement concrete mixtures to which a water-soluble or emulsified polymer has been added. They are known as polymer portland-cement concretes (PPCC). These materials may be formulated to provide improved bonding characteristics, higher strengths, and lower water and chloride permeabilities compared to conventional concrete (ACI 548.1R).

b. Applications and limitations.

(1) Typically, epoxy mortar or concrete is used for overlay thicknesses of about 6 to 25 mm (0.25 to 1 in.). For overlays between 25 and 51 mm (1 and 2 in.) thick, latex-modified concrete is typically used. Conventional portland-cement concrete is typically used in overlays thicker than about 51 mm (2 in.).

(2) Overlays composed of epoxy mortars or concretes are best suited for use in areas where concrete is being attacked by an aggressive substance such as acidic water or some other chemical in the water. These overlays may also be used in some instances to repair surface cracking, provided that the cause of the cracking is well understood and no movement of the concrete is expected in the future. Possible applications for epoxy-based overlays and coatings must be reviewed very carefully to ensure that the proposed use is compatible with the base material. Thermal compatibility is particularly important in exposed repairs that are subjected to wide variations in temperature.

(3) Slab-on-grade or concrete walls with backfill in freezing climates should never receive an overlay or coating that is a vapor barrier. An impervious barrier will cause moisture passing from the subgrade or backfill to accumulate under or behind the barrier, leading to rapid deterioration by cycles of freezing and thawing. A barrier of this type can be a particular problem where the substrate is nonair-entrained concrete subject to cycles of freezing and thawing.

(4) Latex-modified concrete overlays have been used extensively over the past several years for resurfacing bridge decks and other flat surfaces (Ramakrishnan 1992). More recently an epoxy-modified concrete has come into use with the development of an emulsified epoxy. These overlays may be used in lieu of conventional portland-cement concrete overlays and can be placed as thin as 13 mm (1/2 in.). They have excellent bonding characteristics. They require more care and experience than conventional portland-cement overlays. Also, a special two-phase curing requires more time and labor and is described below.

c. Procedure.

(1) Epoxy overlays. Repair of deteriorated concrete with epoxy overlays will involve the use of epoxy concrete or epoxy mortar. Epoxy resin systems conforming to ASTM C 881 (CRD C 595) are suitable.

(a) Generally, aggregates suitable for portland-cement mixtures are suitable for epoxy-resin mixtures. Aggregates are added to the system for economy and improved performance in patching applications and floor toppings. Aggregates should be clean and dry at the time of use and conditioned to a temperature within the range at which the epoxy-resin mortar or concrete is to be mixed. The grading should be uniform with the smallest size passing the No. 100 sieve and the maximum size not to exceed one-third of the mean depth of the patch or opening to be filled. However, the recommended maximum aggregate size for epoxy-resin concrete is 25 mm (1 in.), whereas the maximum size aggregate commonly used for epoxy-resin mortar corresponds to material that will pass a No. 8 sieve.

(b) Aggregates should be used in the amount necessary to ensure complete wetting of the aggregate surfaces. The aggregate-resin proportions will therefore vary with the type and grading of the aggregates. Up to seven parts by weight of the fine aggregate can be mixed with one part of epoxy resin, but a three-to-one proportion is the usual proportion to use for most fine aggregates in making epoxy mortar. For epoxy concrete, the proportion of aggregate to the mixed resin may be as high as 12 to 1 by weight for aggregates in the specific gravity range of 2.50 to 2.80. The aggregate-epoxy proportions also depend on the viscosity of the mixed epoxy system. Since temperature affects the viscosity of the system, the proportions also are dependent on the temperature at which the system is mixed. The trial batches should be made at the temperature of mixing to establish the optimum proportions for the aggregates.

(c) Machine mixing of the epoxy-resin components is mandatory except for mixing volumes of 0.5 L (1 pint) or less. Epoxy mortar or concrete may be machine- or hand-mixed after the epoxy components have been mixed. Small drum mechanical mixers have been used successfully but are difficult to clean properly. Large commercial dough or masonry mortar mixers have been widely and successfully used and present less difficulty in cleaning. Hand-mixing may be performed in metal pans with appropriate tools. When epoxy mortar is hand-mixed, the mixed epoxy system is transferred to the pan, and the fine aggregate is gradually added during mixing. Regardless of how the epoxy concrete is mixed, the fine aggregate is added first and then the coarse aggregate. This procedure permits proper wetting of the fine aggregate particles by the mixed epoxy system and produces a slightly "wet" mixture to which the coarse aggregate is added.

(d) Prior to placement, a single prime coat of epoxy should be worked into the cleaned substrate by brushing, trowelling, or any other method that will thoroughly wet the substrate. The epoxy mortar or concrete must be applied while the prime coat is in a tacky condition. If the depth of the patch is greater than 500 mm (2 in.), placement should be accomplished in lifts or layers of less than 50 mm (2 in.) with some delay between lifts to permit as much heat dissipation as possible. The delay should not extend beyond the setting time of the epoxy formulation. Hand tampers should be used to consolidate the epoxy concrete, taking great care to trowel the mortar or concrete onto the sides and into the corners of the patch. Because of the relatively short pot life of epoxy systems, the placing, consolidating, and finishing operations must be performed without delay.

(e) In final finishing, excess material should not be manipulated onto concrete adjacent to the patch because the carryover material is difficult to clean up. In finishing operations, proper surface smoothness must be achieved. The epoxy mixture tends to build up on the finishing tools, requiring frequent cleaning with an appropriate solvent. After each cleaning, the tool surfaces must be wiped free of excess solvent.

(f) The materials used in the two epoxy systems and the solvents used for cleanup do not ordinarily present a health hazard except to hypersensitive individuals. The materials may be handled safely if adequate precautionary measures are observed. Safety and health precautions for use with epoxies are given in TM 5-822-9, Repair of Rigid Pavements Using Epoxy-Resin Grout, Mortars, and Concrete.

(2) Latex-modified overlays. Styrene-butadiene is the most commonly used latex for concrete overlays (Clear and Chollar 1978).

(a) The materials and mixing procedures for latex-modified mortar and concrete are similar to those for conventional concrete portland-cement mortar and concrete. Latexes in a dispersed form are simply used in larger quantities in comparison to other chemical admixtures. The construction procedure for latex-modified concrete overlays parallels that for conventional concrete overlays except that (1) the mixing equipment must have a means of storing and dispensing the latex into the mixture, (2) the latex-modified concrete has a high slump (typically 125 ± 25 mm (5 ± 1 in.)) and is not air-entrained, and (3) a combination of wet and dry curing is required.

(b) Latex-modified concrete has been produced almost exclusively in mobile, continuous mixers fitted with an additional storage tank for the latex. The latex modifier should always be maintained between 7 and 30 °C (45 and 85 °F). Maintaining the correct temperature may present serious difficulties, especially during the summer months, and may necessitate night placing operations. Hot weather also causes rapid drying of the latex-modified concrete, which promotes shrinkage cracks.

(c) The bond coat consisting of the mortar fraction of the latex-modified concrete is usually produced directly from the continuous mixer by eliminating the coarse aggregate from the mixture. The slurry is broomed into the concrete surface.

(d) Placing operations are straightforward. Finishing machines with conventional vibratory or oscillating screeds may be used, though a rotating cylindrical drum is preferred. Hand finishing is comparable to conventional concrete overlays.

(e) Wet burlap must be applied to the concrete as soon as it will be supported without damage. After 1 to 2 days, the burlap is removed and the overlay should be permitted to air dry for a period of not less than 72 hr. The initial period of wet curing is necessary for the hydration of the portland cement and to prevent the formation of shrinkage cracks; the period of air drying is necessary to permit the latex to dry out and the latex to coalesce and form a continuous film. The film formation within the concrete gives the concrete good bond, flexural strength, and low permeability. The film-forming properties of the latex are temperature sensitive and develop very slowly at temperatures lower than 13 °C (55 °F). Placing and curing should not be done at temperatures lower than 7 °C (45 °F).

(f) See ACI 548.4 for a standard specification for latex-modified concrete overlays. Case histories of repairs with polymer-modified concrete overlays are described by Campbell (1994).

6-17. Overlays (Portland-Cement)

a. Description. Overlays are simply layers of concrete (usually horizontal) placed over a properly prepared existing concrete surface to restore a spalled or disintegrated surface or increase the load-carrying capacity of the underlying concrete. The overlay thickness typically ranges from 102 to 610 mm (4 to 24 in.), depending upon the purpose it is intended to serve. However, overlays as thin as 38 mm (1-1/2 in.) have been placed. For information on polymer-based overlays see Section 6-16.

b. Applications and limitations.

(1) A portland-cement-concrete overlay may be suitable for a wide variety of applications, such as resurfacing spalled or cracked concrete surfaces on bridge decks or lock walls, increasing cover over reinforcing steel, or leveling floors or slabs. Other applications of overlays include repair of concrete surfaces which are damaged by abrasion-erosion and the repair of deteriorated pavements (TM 5-822-6).

(2) Portland-cement-concrete overlays should not be used in applications in which the original damage was caused by aggressive chemical attack that would be expected to act against the portland cement in the overlay. Bonded overlays should not be used in situations in which there is active cracking or structural movement since the existing cracks can be reflected through the overlay or the movement can induce cracks in the overlay; unbonded overlays should be used in these situations.

c. Procedure. The general procedure for applying overlays is as follows: removal of the existing deteriorated concrete; preparation of the concrete surface, including sand- or waterblasting the concrete surface and applying a bonding agent to the surface, if necessary; and placing, consolidating, and curing the overlay. Case histories of repairs with a variety of concrete overlays are described by Campbell (1994).

(1) The guidance given in Chapter 5 should be followed for removal of deteriorated concrete and for preparing the concrete surface. For a bonded overlay to perform properly, the surface to which it is to be bonded must be clean, dry, rough, and dust-free.

(2) The potential for cracking of restrained concrete overlays should be recognized. Any variations in concrete materials, mixture proportions, and construction practices that will minimize shrinkage or reduce concrete temperature differentials should be considered for bonded overlays. Reduced cracking in resurfacing of lock walls has been attributed to lower cement content, larger maximum size coarse aggregate, lower placing and curing temperatures, smaller volumes of placement, and close attention

to curing (Wickersham 1987). Preformed contraction joints 1.5 m (5 ft) on center have been effective in controlling cracking in vertical and horizontal overlays. The critical timing of saw cutting necessary for proper joint preparation is such that this procedure is not recommended for concrete overlays. Where structural considerations permit, an unbonded overlay may be used to minimize cracking caused by restrained contraction of the concrete overlay.

(3) Placing, consolidating, and curing of conventional concrete overlays should follow the guidance given in EM 1110 -2-2000.

6-18. Polymer Coatings

a. Description. Polymer coatings, if the right material for the job condition is selected and properly applied, can be an effective protective coating to help protect the concrete from abrasion, chemical attack, or freeze and thaw damage. Epoxy resins are widely used for concrete coatings. Other polymer coatings include polyester resins and polyurethane resins (ACI 503R and ACI 515.1R).

b. Applications and limitations.

(1) Epoxy resin is used as a protective coating because of its impermeability to water and resistance to chemical attack. It is important that any polymer coating be selected from material designed specifically for the intended application. Some formulations will adhere to damp surfaces and even underwater but many require a completely dry surface. Mixing and applying polymers below 16 °C (60 °F) and above 32 °C (89 °F) will require special caution and procedures. Special sharp sand must be broadcast on the fresh surface if foot traffic is expected on the finished surface. Because of their high exotherm and higher shrinkage values, a neat epoxy in thicker sections is likely to crack.

(2) Slab-on-grade, concrete walls with backfills, or any slab not completely protected from rainwater and subject to freezing and thawing should never receive a coating that will form a vapor barrier. Moisture passing through the subgrade, backfill, or from rain water can accumulate under the coating which will be disrupted by freezing and thawing.

c. Procedure. See applicable portions of Section 6-16.

6-19. Polymer Concrete/Mortar

a. Description. Polymer concrete (PC) is a composite material in which the aggregate is bound together in a dense matrix with a polymer binder (ACI 548.1R). A variety of polymers are being used; the best known and most widely used is epoxy resin (ACI 503R). Some of the other most widely used monomers for PC patching materials include unsaturated polyester resins, a styrene, MMA, and vinylesters. Polymer concrete is quicker setting, has good bond characteristics, good chemical resistance, and high tensile, flexural, and compressive strength compared to conventional concrete. Epoxy resins should meet the requirements of ASTM C 881 (CRD-C 595). The correct type, grade, and class to fit the job should be specified. REMR Technical Note CS-MR-7.1 (USAEWES 1985e) provides general information on eight different types of polymer systems and typical application in maintenance and repair of concrete structures.

b. Applications and limitations.

(1) Epoxy resins can be formulated for a wide range of physical and chemical properties. Some epoxies must be used on dry concrete while others are formulated for use on damp concrete and even underwater. Epoxy hardening is very temperature dependent, and epoxies resins are difficult to apply at temperatures lower than about 16 °C (60 °F). Below 10 °C (50 °F) artificial heating of the material and the substrate must be employed. It is important that epoxy resin or other polymers be selected from material designed specifically for the intended use. Thermosetting polymers, such as polyester and epoxy, exhibit shrinkage during hardening. The shrinkage can be reduced by increasing the amount of aggregate filler.

(2) Other polymers including acrylic polymers (MMA's, HMWM's) and polyesters are being used to make PC. A number of commercial companies now market acrylic-polymer concrete and polyester-polymer concrete used for patching concrete and for overlays. The polyester PC is more widely available because of moderate cost. Polyester resins are more sensitive to moisture than epoxy resins and must be applied on dry concrete.

c. Procedure. Epoxy resins should meet the requirements of ASTM C 881 (CRD-C 595), Type III. For procedures see Section 6-16.

6-20. Polymer Portland-Cement Concrete

a. Description. Polymer portland-cement concrete (PPCC) mixtures are normal portland-cement concrete mixtures to which a water-soluble or emulsified polymer has been added during the mixing process (ACI 548.1R). PPCC has at times been called polymer-modified concrete. The addition of a polymer to portland-cement concrete or mortar can improve strength and adhesive properties. Also, these materials have excellent resistance to damage by freezing and thawing, a high degree of permeability, and improved resistance to chemicals, abrasion, and impact. Latex polymers have been most widely used and accepted. They include styrene butadiene, acrylics, polyvinyl chlorides, and polyvinyl acetates.

b. Applications and limitations. PPCC has superior adhesive properties and can be used in thinner patches and overlays than conventional portland-cement concrete; however, they should not be featheredged. Properties of latexes used in concrete vary considerably so that care should be taken to choose the material best suited for job conditions. Polyvinyl acetates will reemulsify in water and should not be used if the repair will be in continuous contact with water. Ambient temperature can greatly effect the working life for many polymers. PPCC should not be placed at temperatures below 7 °C (45 °F).

c. Procedures. Mixing and handling procedures for PPCC are similar to those used for conventional concrete and mortar; however, curing is different. The film-forming feature of PPCC is such that 1 to 2 days of moist curing followed by air curing is usually sufficient (ACI 548.3R). See Ramakrishnan (1992) for construction practices and specifications for latex-modified concrete.

6-21. Polymer Impregnation

a. Description. Polymer impregnated concrete (PIC) is a portland-cement concrete that is subsequently polymerized (ACI 548.1R). This technique requires use of a monomer system, which is a liquid that consists of small organic molecules capable of combining to form a solid plastic. Monomers have varying degrees of volatility, toxicity, and flammability and do not mix with water. They are very fluid and will soak into dry concrete and fill the cracks. Monomer systems used for impregnation contain a catalyst or initiator and the basic monomer (or different isomers of the same monomer). The systems may also contain a cross-linking agent. When heated, the monomers join together, or polymerize to become a tough, strong, durable plastic, which in concrete greatly enhances a number of the properties of the concrete.

b. Applications and limitations. Polymer impregnation can be used for repair of cracks (ACI 224.1R). If a cracked concrete surface is dried, flooded with the monomer, and polymerized in place, the cracks will be filled and structurally repaired. However, if the cracks contain moisture, the monomer will not soak into the concrete and, consequently, the repair will be unsatisfactory. If a volatile monomer evaporates before polymerization, it will be ineffective. Polymer impregnation has not been used successfully to repair fine cracks. Use of this system requires experienced personnel and some special equipment.

c. Procedure. Badly fractured beams have been repaired with polymer impregnation by drying the fracture, temporarily encasing it in a watertight (monomer proof) band of sheet metal, soaking the fractures with a monomer, and polymerizing the monomer. Large voids or broken areas in compression zones can be filled with fine and coarse aggregate before flooding them with the monomer, providing a polymer-concrete repair. A detailed discussion of polymer impregnation is given in ACI 548.1R. See also the gravity soak procedure described in Section 6-10.

6-22. Polymer Injection

a. Description. Polymers commonly used to repair cracks or joints by injection may be generally categorized as either rigid or flexible systems. Epoxies are the most common rigid systems used for structural repair or "welding" of cracks to form a monolithic structure. Flexible polyurethane systems are most often used for stopping water flow and sealing active cracks. Cracks as narrow as 0.05 mm (0.002 in.) can be bonded by the injection of epoxy (ACI 224.1R). The technique generally consists of drilling holes at close intervals along the cracks, in some cases installing entry ports, and injecting the epoxy under pressure. Although the majority of the injection projects have been accomplished with high-pressure injection, some successful work has been done with low pressures.

b. Applications and limitations.

(1) Rigid repairs. Epoxy injection has been successfully used in the repair of cracks in buildings, bridges, dams, and other types of concrete structures. However, unless the crack is dormant (or the cause of cracking is removed, thereby making the crack dormant), cracking will probably recur, and structural repair by injection should not be used. With the exception of certain specialized epoxies, this technique is not applicable if the cracks are actively leaking and cannot be dried out. While moist

cracks can be injected, contaminants in the crack (including water) will reduce the effectiveness of the epoxy to structurally repair the crack. Epoxy injection can also be used in the repair of delaminations in bridge decks.

(2) Flexible repairs. If the cracks are active and it is desired to seal them while allowing continued movement at these locations, it is necessary to use a grout that allows the filled crack to act as a joint. This is accomplished by using a polymer which cures into a closed-cell foam. Water-activated polyurethane grouts, both hydrophobic and hydrophilic, are commonly used for sealing leaking cracks. Solomon and Jaques (1994) provide an excellent discussion of materials and methods for injecting leaking cracks. Applications of water-activated polyurethanes in repair of waterstop failures are discussed in Section 8-2. Also, see Section 6-11.

(3) Polymer injection generally requires a high degree of skill for satisfactory execution, and application of the technique may be limited by ambient temperature.

c. *High-pressure injection procedure.* The majority of injection projects are accomplished with high-pressure injection (350 KPa (50 psi) or higher). The general steps involved in epoxy injection are as follows (ACI 224.1R).

(1) Clean the cracks. The first step is to clean the cracks that have been contaminated. Oil, grease, dirt, or fine particles of concrete prevent epoxy penetration and bonding. Preferably, contamination should be removed by flushing with water or, if the crack is dry, some other specially effective solvent. The solvent is then blown out with compressed air, or adequate time is allowed for air drying.

(2) Seal the surfaces. Surface cracks should be sealed to keep the polymer from leaking out before it has gelled. Where the crack face cannot be reached but where there is backfill or where a slab-on-grade is being repaired, the backfill material or subbase material is often an adequate seal. A surface can be sealed by brushing an epoxy along the surface of the crack and allowing it to harden. If extremely high injection pressures are needed, the crack should be cut out to a depth of 13 mm (1/2 in.) and width of about 20 mm (3/4 in.) in a V-shape, filled with an epoxy, and struck off flush with the surface. If a permanent glossy appearance along the crack is objectionable and if high injection pressure is not required, a strippable plastic may be applied along the crack.

(3) Install the entry ports. Three methods are in general use:

(a) Drilled holes--fittings inserted. Historically, this method was the first to be used and is often used in conjunction with V-grooving of the cracks. The method entails drilling a hole into the crack, approximately 19 mm (3/4 in.) in diam and 13 to 25 mm (1/2 to 1 in.) below the apex of the V-grooved section, into which a fitting such as a pipe nipple or tire valve stem is bonded with an epoxy adhesive. A vacuum chuck and bit are useful in preventing the cracks from being plugged with drilling dust. Hydrostatic pressure tests showed that molded injection ports mounted within a drilled port hole can withstand pressures of 1.4 to 1.9 MPa (200 to 275 psi) before leaks begin to develop. In comparison, surface-mounted ports withstood pressures between 0.3 and 1.0 MPa (50 and 150 psi), depending on the type of port (Webster, Kukacka, and Elling 1990).

(b) Bonded flush fitting. When the cracks are not V-grooved, a method frequently used to provide an entry port is to bond a fitting flush with the concrete face over the crack. This flush fitting has a hat-like cross section with an opening at the top for the adhesive to enter.

(c) Interruption in seal. Another system of providing entry is to omit the seal from a portion of the crack. This method can be used when special gasket devices are available that cover the unsealed portion of the crack and allow injection of the adhesive directly into the crack without leaking.

(4) Mix the epoxy. Epoxy systems should conform to ASTM C 881 (CRD-C 595), Type I, low-viscosity grade. Mixing is done either by batch or continuous methods. In batch mixing, the adhesive components are premixed according to the manufacturer's instructions, usually with the use of a mechanical stirrer, such as a paint-mixing paddle. Care must be taken to mix only the amount of adhesive that can be used prior to commencement of gelling of the material. When the adhesive material begins to gel, its flow characteristics begin to change, and pressure injection becomes more and more difficult. In the continuous mixing system, the two liquid adhesive components pass through metering and driving pumps prior to passing through an automatic mixing head. The continuous mixing system allows the use of fast-setting adhesives that have a short working life.

(5) Inject the epoxy. Hydraulic pumps, paint pressure pots, or air actuated caulking guns can be used. The pressure used for injection must be carefully selected. Increased pressure often does little to accelerate the rate of injection. In fact, the use of excessive pressure can propagate the existing cracks, causing additional damage.

If the crack is vertical, the injection process should begin by pumping into the entry port at the lowest elevation until the epoxy level reaches the entry port above. The lower injection port is then capped, and the process is repeated at successively higher ports until the crack has been completely filled and all ports have been capped. For horizontal cracks, the injection should proceed from one end of the crack to the other in the same manner. The crack is full if the pressure can be maintained. If the pressure cannot be maintained, the epoxy is still flowing into unfilled portions or leaking out of the crack.

(6) Remove the surface seal. After the injected epoxy has cured, the surface seal should be removed by grinding or other means, as appropriate. Fittings and holes at entry ports should be painted with an epoxy patching compound.

d. Alternate high-pressure procedure. To develop alternatives to concrete removal and replacement in repair of mass concrete hydraulic structures, a study was initiated, as part of the REMR Research Program, to evaluate in situ repair procedures.

(1) Eight injection adhesives were experimentally evaluated to determine their effectiveness in the repair of air-dried and water-saturated cracked concrete. The adhesives were three epoxies, an emulsifiable polyester resin, furfuryl alcohol, a furan resin, a high-molecular-weight methacrylate, and a polyurethane. Because of their low bond strength to water-saturated concrete, the furan resin, furfuryl alcohol, and the polyurethane were not considered further as injection adhesives. The remaining adhesives were used to repair both air-dried and water-saturated concrete slabs by conventional injection. The most promising adhesive was a two-component, very low-viscosity epoxy system designed specifically for pressure injection repairs (Webster and Kukacka 1988).

(2) A field test was performed on a tainter gate pier stem at Dam 20, Mississippi River, to demonstrate, under actual field conditions, the procedures developed in the laboratory and to evaluate the effectiveness of the materials and equipment selected for use (Webster, Kukacka, and Elling 1989). Problem areas identified during the field test were addressed in development of a modified repair procedure. Modifications included a better method for attaching the injection ports to the concrete and drilling small-diameter holes into the concrete to facilitate epoxy penetration into the multiple, interconnecting cracks. The modified procedure was demonstrated at Dam 13 on the Mississippi River near Fulton, Illinois (Webster, Kukacka, and Elling 1990).

(3) The first step in this repair procedure is to clean the concrete surfaces by sandblasting. Next, injection holes are drilled. These holes, 13 mm in (1/2 in.) diam and 152 m (6 in.) deep, are wet drilled to flush fines from the holes as they occur. After injection ports are installed, the entire surface of the repair area is sealed with epoxy. After the seal has cured, injection is begun.

(4) Visual examinations of cores taken after injection indicate that a crack network within 152 to 254 mm (6 to 10 in.) of the surface can be filled with epoxy. These examinations indicate that the special injection procedure works very well and laboratory tests substantiate this conclusion. For example, splitting tensile strengths of the repaired cores average more than twice that of the unrepaired cores and only 10 percent less than the strength of the uncracked concrete.

e. Low-pressure injection. Similar results are attainable with either low-pressure or high-pressure injection procedures. For example, results achieved through an injection pressure of 2 MPa (300 psi) for 3 min are reportedly duplicated at a pressure of only 0.03 MPa (5 psi) or less for a period of 1 hr, presuming a low-viscosity, long pot life resin is used (Trout 1994). Generally, anything that can be injected with high pressure can be injected with low pressure; it just takes longer, which accounts for the selection of high-pressure systems for most large projects. However, there are situations where low-pressure injection has distinct advantages.

(1) Low injection pressures allow the use of easily removable materials for sealing the surface of the crack, whereas high-pressure injection normally requires an epoxy seal and aggressive removal procedures. Seals that are easily removed minimize the potential for surface blemishes which is particularly important for architectural concrete. Some units designed specifically for low pressure use can maintain pressures of less than 0.01 MPa (1 psi) for delicate projects such as repair of murals and mosaics.

(2) Low-pressure systems are portable, easy to mobilize, require little support from other construction equipment, and their initial cost is about one-tenth the cost of a high-pressure system.

(3) Low-pressure injection is less hazardous, and the use of skilled or experienced labor is seldom critical. Typically, low-pressure systems use prebatched resin rather than metering dispensers. Once the resin is mixed, it is pressurized by air or springs within capsules, inflatable syringe-like devices, that are left in place until

the resin has gelled. The use of long pot life resins is essential for successful low-pressure injection: a gel time of 1 hr at 22 °C (72 °F) is recommended.

6-23. Precast Concrete

a. Description. Precast concrete is concrete cast elsewhere than its final position. The use of precast concrete in repair and replacement of structures has increased significantly in recent years and the trend is expected to continue. Compared with cast-in-place concrete, precasting offers a number of advantages including ease of construction, rapid construction, high quality, durability, and economy.

b. Applications and limitations. Typical applications of precast concrete in repair or replacement of civil works structures include navigation locks, dams, channels, floodwalls, levees, coastal structures, marine structures, bridges, culverts, tunnels, retaining walls, noise barriers, and highway pavement.

c. Procedures. Procedures for use of precast concrete in repair of a wide variety of structures are described in detail by McDonald and Curtis (in preparation). Case histories describing the use of precast concrete in repair of navigation lock walls are described in Section 8-1. Selected case histories of additional precast concrete applications are summarized in Section 8-5.

6-24. Preplaced-Aggregate Concrete

a. Description. Preplaced-aggregate concrete is produced by placing coarse aggregate in a form and then later injecting a portland-cement-sand grout, usually with admixtures, to fill the voids. As the grout is pumped into the forms, it will fill the voids, displacing any water, and form a concrete mass.

b. Applications and limitations. Typically, preplaced-aggregate concrete is used on large repair projects, particularly where underwater concrete placement is required or when conventional placing of concrete would be difficult. Typical applications have included underwater repair of stilling basins, bridge piers, abutments, and footings. Applications of preplaced-aggregate concrete in repair of navigation lock walls are described in Section 8-1. The advantages of using preplaced-aggregate concrete include low shrinkage because of the point-to-point aggregate contact, ability to displace water from forms as the grout is being placed, and the capability to work around a large number of blockouts in the placement area.

c. Procedure. Guidance on materials, mixture proportioning, and construction procedures for preplaced-aggregate concrete can be found in EM 1110-2-2000 and in ACI 304.1R.

6-25. Rapid-Hardening Cements

a. Description. Rapid-hardening cements are defined as those that can develop a minimum compressive strength of 20 MPa (3,000 psi) within 8 hr or less. The types of rapid-hardening cements and patching materials available and their properties are described in REMR Technical Note CS-MR-7.3 (USAEWES 1985g). A specification for prepackaged, dry, rapid-hardening materials is given in ASTM C 928.

b. Applications and limitations.

(1) Magnesium-phosphate cement (MPC). This material can attain a compressive strength of several thousand pounds per square inch in 1 hr. MPC is useful for cold-weather embedments and anchoring and for patching applications where a short downtime can justify the additional expense. Finishing must be performed quickly because of the rapid set. MPC must be used with non-calcareous aggregates. MPC has low, long-term shrinkage and is nonreactive to sulphates. MPC is air-cured in a manner similar to the way epoxy concrete is cured. A damp substrate will adversely affect hardening.

(2) High alumina cements (HAC). The 24-hr strength of HAC is approximately equivalent to the 28-day strength of portland-cement concrete. The initial set however is reported to be up to 3 hr, which may be beneficial for transportation of the mixed concrete. HAC is more stable at high temperature than portland cement, providing aggregates that resist the high temperatures are used. A disadvantage is that when high alumina cement is subjected to in-service conditions of high humidity and elevated temperatures greater than 20 °C (68 °F) there is a "conversion reaction" which can cause a drastic strength loss (Mailvaganam 1992).

(3) Regulated-set portland cement. The initial set time is 15-20 min, but the set may be retarded by the use of citric acid. Regulated-set portland cement is not recommended for use in concrete exposed to sulphate soils or water.

(4) Gypsum cements. Gypsum cements are fast-setting and can obtain compressive strengths of as much as 21 m MPa (3,000 psi) in 30 min. For the most part, however, they are not as durable as portland-cement

concrete. They abrade easily, are not as frost resistant, and may be affected by fuel or solvent spills.

(5) Special blended cements. There are many different types of blended cements available. These materials generally have very high-early strengths, and setting times may be adjusted so that they may be transported by ready-mix truck.

(6) Packaged patching materials. There are numerous rapid-hardening patching materials available from different suppliers. Many are excellent materials for a variety of uses, although the claims of certain attributes by some suppliers have not been borne out by testing. ASTM C 928 is a specification that can be used for these materials; however, this specification does not provide requirements for bond strength, for freeze-thaw durability, for sulphate exposure or alkali reactivity. These materials should be used only when a service record for the proposed material, in the same environment, is available or when government testing is performed.

c. Procedure. These materials should be mixed and placed in accordance with the suppliers recommendations.

6-26. Roller-Compacted Concrete

a. Description. Roller-compacted concrete (RCC) is defined as "concrete compacted by roller compaction; concrete that, in its unhardened state, will support a roller while being compacted" (ACI 116R). Properties of hardened RCC are similar to those of conventionally placed concrete.

b. Applications and limitations. RCC should be considered where no-slump concrete can be transported,

placed, and compacted with earth and rock-fill construction equipment. Ideal RCC projects will involve large placement areas, little or no reinforcement or embedded metals, or other discontinuities such as piles.

(1) The primary applications of RCC within the Corps of Engineers have been in new construction of dams and pavement. Meanwhile, RCC has been so successful for repair of non-Corps dams that the number of dam repair projects now exceeds the number of new RCC dams. The primary advantages of RCC are low cost (25 to 50 percent less than conventionally placed concrete) and rapid construction.

(2) RCC has been used to strengthen and improve the stability of existing dams, to repair damaged overflow structures, to protect embankment dams during overtopping, and to raise the crest on existing dams. Selected applications of RCC in repair of a variety of structures are summarized in Section 8-8.

c. Procedures. Guidance on the use of RCC is given in EM 1110-2-2006 and ACI 207.5R.

6-27. Routing and Sealing

a. Description. This method involves enlarging the crack along its exposed face and filling and sealing it with a suitable material (Figure 6-10). The routing operation may be omitted but at some sacrifice in the permanence of the repair. This is the simplest and most common method for sealing dormant cracks.

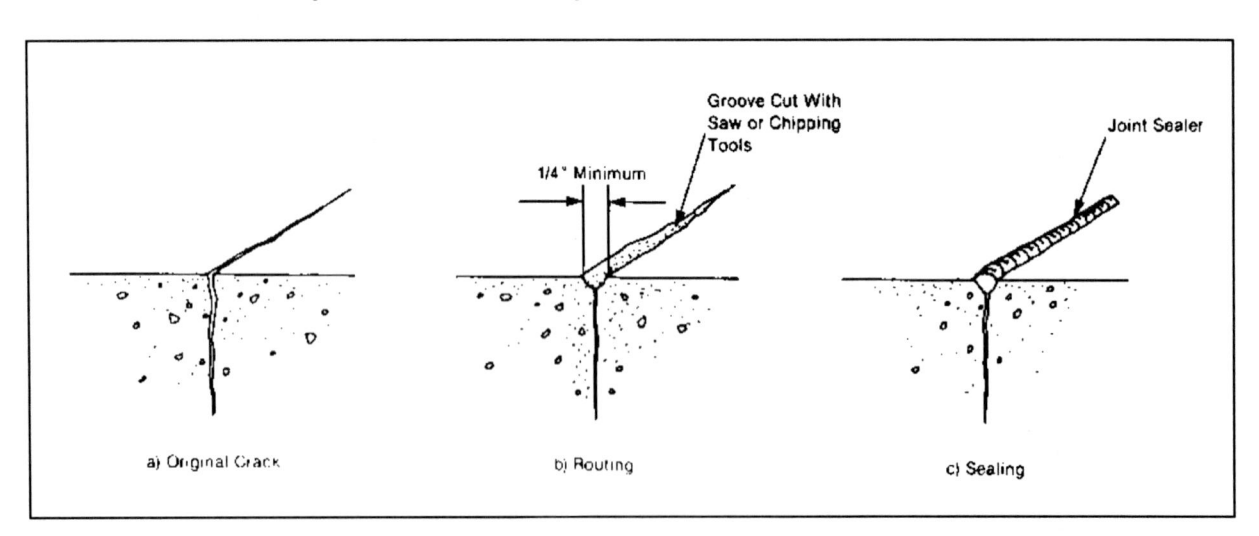

Figure 6-10. Repair of crack by routing and sealing

b. Applications and limitations. This method can be used on cracks that are dormant and of no structural significance. It is applicable to sealing both fine pattern cracks and large isolated defects. It will not be effective in repair of active cracks or cracks subject to significant hydrostatic pressure. However, some reduction in flow may be obtained when this method is used to seal the pressure face of cracks subject to hydrostatic pressure.

c. Procedure.

(1) The routing operation consists of following along the crack with a concrete saw or with hand or pneumatic tools and opening the crack sufficiently to receive the sealant. A minimum surface width of 6 mm (1/4 in.) is desirable since smaller openings are difficult to fill. The surfaces of the routed joint should be cleaned and permitted to dry before sealing.

(2) The purpose of the sealant is to prevent water from reaching the reinforcing steel, hydrostatic pressure from developing within the joint, the concrete surface from staining, or moisture problems on the far side of the member from developing. The sealant may be any of several materials, depending on how tight or permanent a seal is desired. Epoxy compounds are often used. Hot-poured joint sealant works very well when thorough watertightness of the joint is not required and appearance is not important. Urethanes, which remain flexible through large temperature variations, have been used successfully in cracks up to 19 mm (3/4 in.) in width and of considerable depth. There are many commercial products, and the manufacturers should be consulted to ascertain the type and grade most applicable to the specific purpose and condition of exposure. The Repair Materials Database (Section 4-5) contains information on a variety of crack repair materials. The method of placing the sealant depends on the material to be used, and the techniques recommended in ACI 504R should be followed.

6-28. Shotcrete

a. Description. Shotcrete is mortar pneumatically projected at high velocity onto a surface. Shotcrete can contain coarse aggregate, fibers, and admixtures. Properly applied shotcrete is a structurally adequate and durable repair material that is capable of excellent bond with existing concrete or other construction materials (ACI 506R).

b. Applications and limitations. Shotcrete has been used to repair deteriorated concrete bridges, buildings, lock walls, dams, and other hydraulic structures. The performance of shotcrete repair has generally been good. However, there are some instances of poor performance. Major causes of poor performance include inadequate preparation of the old surface and poor application techniques by inexperienced personnel. Satisfactory shotcrete repair is contingent upon proper surface treatment of old surfaces to which the shotcrete is being applied. In a repair project where thin repair sections (less than 150 m (6 in.) deep) and large surface areas with irregular contours are involved, shotcrete is generally more economical than conventional concrete because of the saving in forming costs. One of the problems in the shotcrete repair is overrun in estimated quantities. These overruns are usually related to underestimating the quantity of deteriorated concrete to be removed. Estimation errors can be minimized by a thorough condition survey as close as possible to the time that the repair work is to be executed. Most shotcrete mixtures have a high cement and therefore a greater potential for drying shrinkage cracking compared to conventional concrete (ACI 506R). Also, the overall quality is sensitive to the quality of workmanship. Problems associated with shotcrete repairs on nonair-entrained concrete are discussed in Section 8-1b.

c. Procedure. Guidance on the selection, proportioning, and application of shotcrete is given in EM 1110-2-2005. In addition, a small hand-held funnel gun was developed by the U.S. Army Engineer Division, Missouri River (1974), for pneumatic application of portland-cement mortar. The gun (Figure 6-11) is easily assembled from readily available material, has only a few critical dimensions, and can be operated by personnel without extensive training. The gun has been used successfully for application of mortar in small, shallow repairs on vertical and overhead surfaces.

6-29. Shrinkage-Compensating Concrete

a. Description. Shrinkage-compensating concrete is an expansive cement concrete which is used to minimize cracking caused by drying shrinkage in concrete slabs, pavements, and structures. Type K, Type M, or Type S expansive portland cements is used to produce shrinkage-compensating concrete. Shrinkage-compensating concrete will increase in volume after setting and during hardening. When properly restrained by reinforcement, expansion will induce tension in the reinforcement and compression in the concrete. On subsequent drying, the shrinkage so produced, instead of causing tensile cracking merely relieves the strains caused by the initial expansion (Figure 6-12).

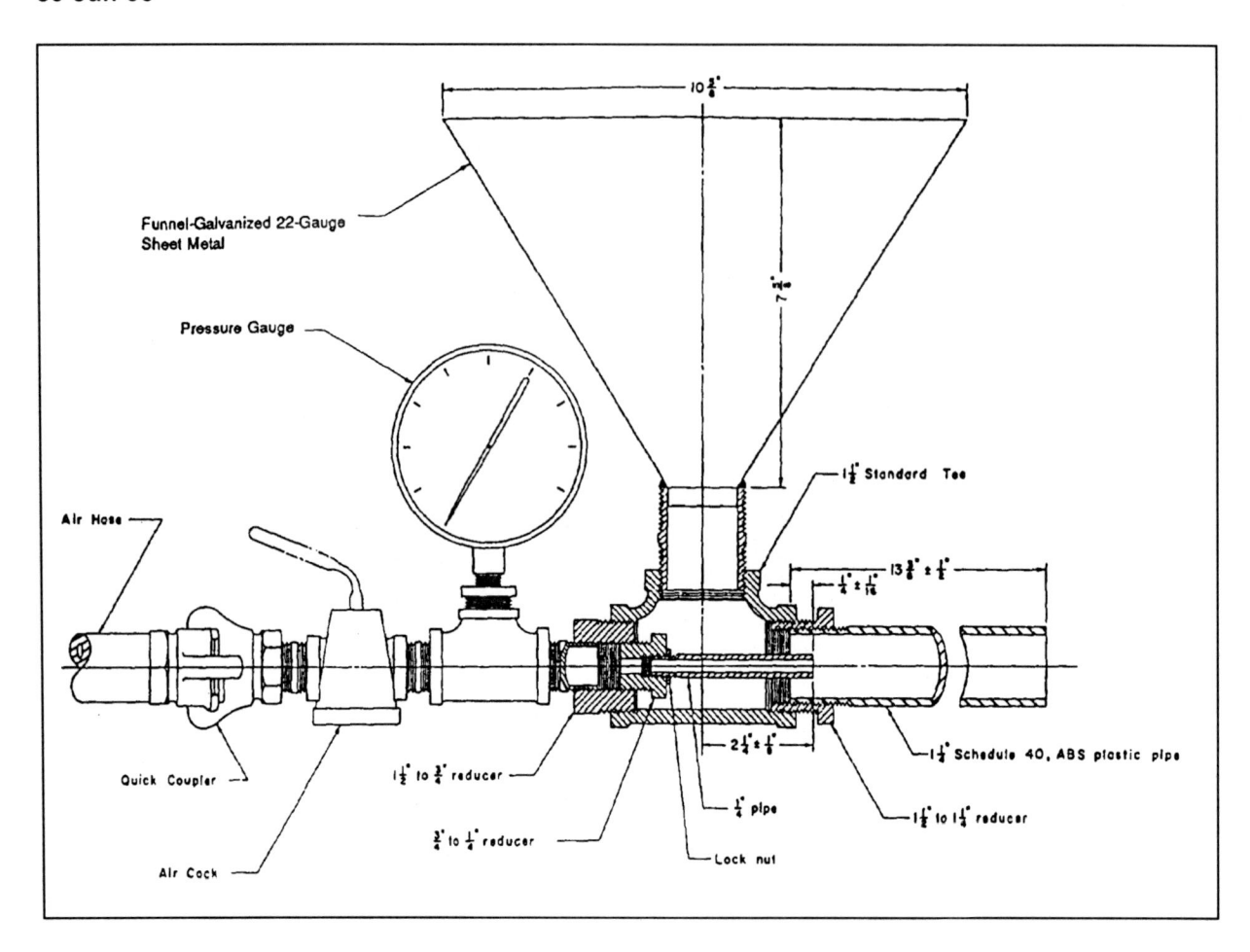

Figure 6-11. Mortar gun for concrete repair (U.S. Army Engineer Division, Missouri River 1974)

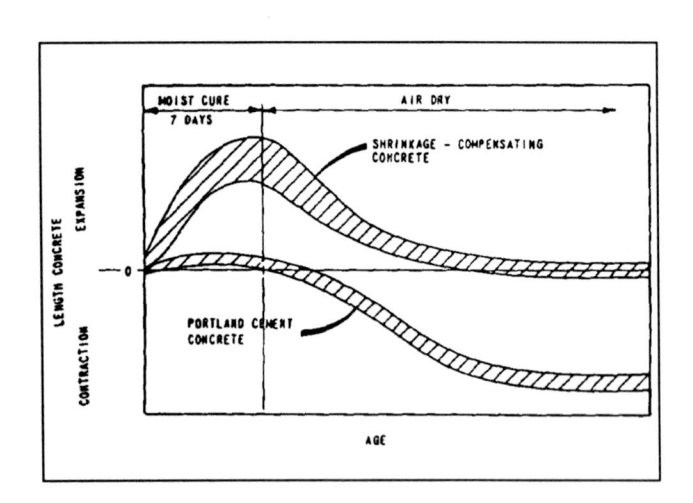

Figure 6-12. Typical length change characteristics of shrinkage-compensating and portland-cement concretes (ACI 223)

b. Applications and limitations. Shrinkage-compensating concrete may be used as bonded or unbonded topping over a deteriorated or cracked concrete slab. The proper amount of internal reinforcement must be provided to develop shrinkage compensation. Early curing and proper curing are very important. Some shrinkage-compensating concrete mixtures will show early stiffening anda loss of workability. It is important to maintain close control over the amount of added mixture water so that the maximum w/c is not exceeded. Some ASTM C 494, Types A, D, F and G admixtures are not compatible with shrinkage-compensating cements. Larger distances may be used between contraction joints. For exposed areas, a maximum of 31 m (100 ft) is recommended. For areas protected from extreme fluctuations in temperature and moisture, joint spacing of 46 to 60 m (150 to 200 ft) have been used.

c. Procedures. Construction with shrinkage-compensating concrete generally follows the precepts for conventional portland-cement concrete. An excellent reference for design using shrinkage-compensating concrete is ACI 223.

6-30. Silica-Fume Concrete

a. Description. Silica fume, a by-product of silicon or ferrosilicon production, is a very fine powder with a medium to dark gray color. The spherical silica-fume particles are typically about 100 times smaller than portland-cement grains. The resulting high surface area is reflected in an increased water demand which can be overcome with a WRA or HRWRA. Silica fume is available in several forms: loose powder, densified powder, slurry, and, in some areas, as a blended portland-silica-fume cement. Silica fume is generally proportioned as an addition, by mass, to the cementitious materials and not as a substitution for any of these materials. The optimum silica-fume content ranges from about 5 to 15 percent by mass of cement. When properly used, silica fume can enhance certain properties of both fresh and hardened concrete, including cohesiveness, strength, and durability. Apparently, concretes benefit from both the pozzolanic properties of silica fume and the extremely small particle size. ACI 226 (1987) provides a detailed discussion on the use of silica fume in concrete.

b. Applications and limitations. The use of silica fume as a pozzolan in concrete produced in the United States has increased in recent years. Silica-fume concrete is appropriate for concrete applications which require very high strength, high abrasion-erosion resistance, very low permeability, or where very cohesive mixtures are needed to avoid segregation (EM 1110-2-2000). Silica-fume concrete should be considered for repair of structures subjected to abrasion-erosion damage, particularly in those areas where locally available aggregate might not otherwise be acceptable.

(1) Silica-fume concrete has been successfully used by the Corps of Engineers in repair of abrasion-erosion damaged concrete in stilling basins (Section 8-3*d*) and channels (Holland and Gutschow 1987). Although the placements generally went well, the silica-fume concrete overlay used to repair the Kinzua Dam stilling basin exhibited extensive cracking. However, these fine cracks have not adversely affected the performance of the repair.

(2) Concrete materials and mixture proportions similar to those used in the stilling basin repair were later used in laboratory tests to determine those properties of silica-fume concrete which might affect cracking (McDonald 1991). None of the material properties, with the possible exception of autogenous volume change, indicated that silica-fume concrete should be significantly more susceptible to cracking as a result of restrained contraction than conventional concrete. In fact, some material properties, particularly ultimate tensile strain capacity, would indicate that silica-fume concrete should have a reduced potential for cracking.

c. Procedure. Silica-fume concrete requires no significant changes from normal transporting, placing, and consolidating practices. However, special considerations in finishing and curing practices may be required as discussed in EM 1110-2-2000. The potential for cracking of restrained concrete overlays, with or without silica fume, should be recognized. Any variations in concrete materials, mixture proportions, and construction practices that will minimize shrinkage or reduce concrete temperature differentials should be considered. Where structural considerations permit, a bond breaker at the interface between the replacement and existing concrete is recommended.

6-31. Slabjacking

a. Description. Slabjacking is a repair process in which holes are drilled in an existing concrete slab and a cementitious grout is injected to fill any voids and raise the slab as necessary. This process is also known as mudjacking.

b. Applications and limitations. Slabjacking is applicable to any situation in which a slab or other concrete section or grade needs to be repositioned. Slabjacking should be considered as an alternative to removal and replacement with conventional concrete. Reported applications include sidewalks, pavement slabs, water tanks, and swimming pools. This process has also been used to fill voids behind and under concrete structures; in such applications, it is simply a variation of portland-cement grouting.

c. Procedure. Information on procedures, materials, and equipment for slabjacking can be found in EM 1110-2-3506 and Meyers 1994.

6-32. Stitching

a. Description. This method involves drilling holes on both sides of the crack and grouting in stitching dogs (U-shaped metal units with short legs) that span the crack (Johnson 1965) (Figure 6-13).

b. Applications and limitations. Stitching may be used when tensile strength must be reestablished across major cracks. Stitching a crack tends to stiffen the structure, and the stiffening may accentuate the overall structural restraint, causing the concrete to crack elsewhere. Therefore, it may be necessary to strengthen the adjacent

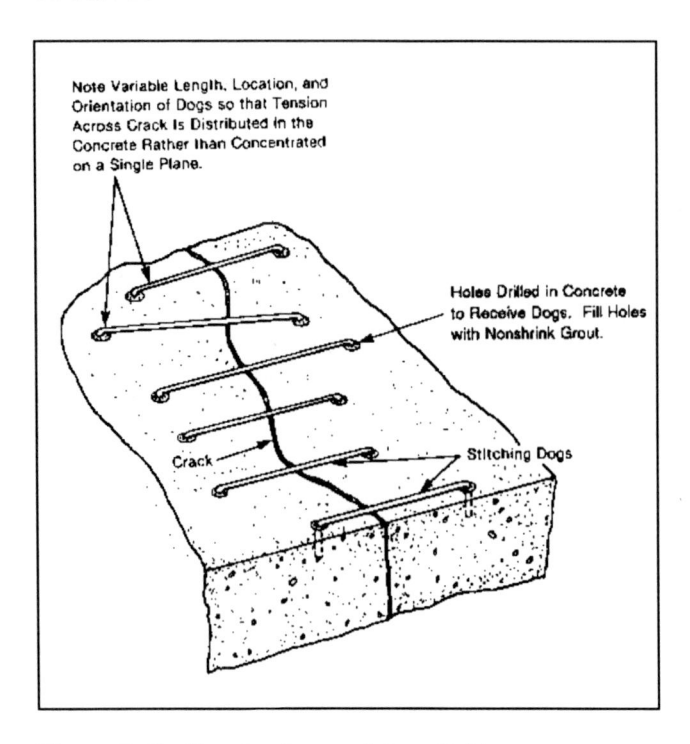

Note Variable Length, Location, and Orientation of Dogs so that Tension Across Crack Is Distributed in the Concrete Rather than Concentrated on a Single Plane.

Holes Drilled in Concrete to Receive Dogs. Fill Holes with Nonshrink Grout.

Crack

Stitching Dogs

Figure 6-13. Repair of a crack by stitching

section with external reinforcement embedded in a suitable overlay.

c. Procedure.

(1) The stitching procedure consists of drilling holes on both sides of the crack, cleaning the holes, and anchoring the legs of the dogs in the holes, with either a nonshrink grout or an epoxy-resin-based bonding system. The stitching dogs should be variable in length and orientation or both, and they should be located so that the tension transmitted across the crack is not applied to a single plane within the section but is spread over an area.

(2) Spacing of the stitching dogs should be reduced at the end of cracks. In addition, consideration should be given to drilling a hole at each end of the crack to blunt it and relieve the concentration of stress.

(3) Where possible, both sides of the concrete section should be stitched so that further movement of the structure will not pry or bend the dogs. In bending members, it is possible to stitch one side of the crack only. Stitching should be done on the tension face, where movement is occurring. If the member is in a state of axial tension, then the dogs must be placed symmetrically, even if excavation or demolition is required to gain access to opposite sides of the section.

(4) Stitching will not close a crack but can prevent it from propagating further. Where there is a water problem, the crack should be made watertight as well as stitched to protect the dogs from corrosion. This repair should be completed before stitching begins. In the case of active cracks, the flexible sealing method (Section 6-9) may be used in conjunction with the stitching techniques.

(5) The dogs are relatively thin and long and cannot take much compressive force. Accordingly, if there is a tendency for the crack to close as well as to open, the dogs must be stiffened and strengthened, for example, by encasement in an overlay.

6-33. Underwater Concrete Placement

a. Description. Underwater concrete placement is simply placing fresh concrete underwater with a number of well recognized techniques and precautions to ensure the integrity of the concrete in place. Concrete is typically placed underwater by use of a tremie or a pump. The quality of cost-in-place concrete can be enhanced by the addition of an antiwashout admixture which increases the cohesiveness of the concrete. The special case in which the concrete is actually manufactured underwater, the preplaced-aggregate technique, is described in Section 6-24. Flat and durable concrete surfaces with in-place strengths and densities essentially the same as those of concrete cast and consolidated above water can be obtained with proper mixture proportioning and underwater placement procedures.

b. Applicability and limitations.

(1) Placing concrete underwater is a suitable repair method for filling voids around and under concrete structures. Voids ranging from a few cubic yards to thousands of cubic yards have been filled with tremie concrete. Concrete pumped underwater or placed by tremie has also been used to repair abrasion-erosion damage on several structures (McDonald 1980). Another specialized use of concrete placed underwater is in the construction of a positive cutoff wall through an earthfill dam. This process is discussed in Section 8-4.

(2) There are two significant limitations on the use of concrete placed underwater. First, the flow of water through the placement site should be minimized while the concrete is being placed and is gaining enough strength to resist being washed out of place or segregated. One approach that may be used to protect small areas is to use top form plates under which concrete may be pumped. The designer, contractor, and inspectors must all be

thoroughly familiar with underwater placements. Placing concrete underwater is not a procedure that all contractors and inspectors are routinely familiar with since it is not done as frequently as other placement techniques. The only way to prevent problems and to ensure a successful placement is to review, in detail, all aspects of the placement (concrete proportions, placing equipment, placing procedures, and inspection plans) well before commencing the placement.

c. Procedures. Guidance on proportioning concrete mixtures for underwater placement is given in EM 1110-2-2000. ACI 304R provides additional information on concrete placed underwater. Underwater repair of concrete is discussed in Section 8-6.

Chapter 7
Maintenance of Concrete

7-1. General

Preventing concrete deterioration is much easier and more economical than repairing deteriorated concrete. Preventing concrete deterioration should actually begin with the selection of proper materials, mixture proportions, and placement and curing procedures. If additional protection against deterioration is required, the need should be recognized and provided for during design of the structure. Of course, all potential hazards to concrete cannot always be predicted, and some well-engineered techniques and procedures may prove unsuccessful. Thus, there is generally a need for follow-up maintenance action. The primary types of maintenance for concrete include timely repair of cracks and spalls, cleaning of concrete to remove unsightly material, surface protection, and joint restoration. Materials and procedures for repair of concrete cracking and spalling have been described in previous chapters. Materials and procedures appropriate for cleaning and protecting concrete surfaces and joint maintenance are described in the following.

7-2. Cleaning

Stains seldom affect the service life of a structure, although they are often unsightly, especially on architectural concrete finishes. Some of the more common stains are iron rust, oil, grease, dirt, mildew, asphalt, efflorescence, soot, and graffiti. Stains often penetrate the exposed surface because concrete is porous and absorbent. Therefore, stains should be removed as soon as possible to prevent deeper migration into the concrete. Also, stains tend to bind more tightly to the concrete with time, and some undergo chemical changes that make removal more difficult. Almost all stains can be removed if the type of stain can be identified and the correct removal method is selected (REMR Technical Note CS-MR-4.4 (USAEWES 1985d)).

   a. Identification. The first step in the removal process is to identify the stain and then select a cleaning agent and method accordingly. If the stain is impossible to identify, potential cleaning materials should be tested in an inconspicuous area in the following order: organic solvents, oxidizing bleaches, reducing bleaches, and acids.

   b. Stain removal. Stains can be removed with several methods including brushing and washing, steam cleaning, water blasting, abrasive blasting, flame cleaning, mechanical cleaning, and chemical cleaning (*Concrete Repair Digest* 1993). Since there is usually more than one method that can be used to remove a given stain, the advantages and limitations of each potential method should be considered in making a final selection.

   (1) Removal methods.

   (a) Water washing. A fine mist spray is recommended, as excessive water pressure can drive the stain farther into the concrete. Washing should be done from the top of the structure down. If the water alone is not cleaning the concrete, it can be used in conjunction with the following in the order listed: a soft brush, a mild soap, a stronger soap, ammonia, or vinegar.

   (b) Steam cleaning. Steam is generally good for removing dirt and chewing gum; however, in most applications it is relatively expensive.

   (c) Water blasting. Water blasting removes less surface material than sandblasting because no abrasive is used; however, a test section to determine the effect of this method on surface texture is recommended.

   (d) Abrasive blasting. Abrasive blasting tends to remove some of the concrete resulting in a nonuniform surface. The nozzle should be held farther from the surface than normal in any kind of blasting to minimize abrasion.

   (e) Flame cleaning. Flame cleaning will remove organic materials that do not respond to solvents. However, this method can cause scaling of the concrete surface and may produce objectionable fumes.

   (f) Mechanical cleaning. Power tools (grinders, buffers, chisels, brushes) may be required to remove the more stubborn stains from concrete. These tools can damage thin sections or remove more concrete than is desirable. Chiselling or grinding can be an effective cleaning method provided a roughened or uneven surface is acceptable.

   (g) Chemical cleaning. Organic solvents can usually be used with little dilution. Inorganic solvents such as ammonium hydroxide, sodium hypochlorite, and hydrogen peroxide can be purchased in ready-mixed solutions; other organic solvents can be purchased as solids and then mixed with water according to manufacturer's directions. It may be desirable to mix the solvent to be used with an inert fine powder to form a poultice which is then troweled over the stain (REMR Technical Note CS-MR-4.4 (USAEWES 1985d)). Chemical cleaning is often the best way to remove stains because most chemicals do not alter the surface texture of the concrete nor do they require the equipment needed by mechanical methods. However, there are safety considerations: many chemicals are mild and safe if used with care, while others are toxic, flammable, or corrosive to concrete. Manufacturer's directions and recommendations for the protection of occupational health and safety should be carefully followed. Material Safety Data Sheets (MSDS) should be obtained from the manufacturers of such materials. In cases where the effects of a chemical substance on occupational health and safety are unknown, chemical substances should be treated as potentially hazardous or toxic materials.

(2) Removing specific stains. Detailed procedures for removing a variety of stains are described in REMR Technical Notes CS-MR-4.3 and 4.4 (USAEWES 1985c and d) and *Concrete Repair Digest* (1993). The procedures are summarized in the following paragraphs.

(a) Iron rust. If the stain is light or shallow, mop the surface with a solution of oxalic acid and water. Wait 2 or 3 hr, and then scrub the surface with stiff brushes while rinsing with clear water. If the stain is deep, prepare a poultice by mixing sodium citrate, glycerol, and diatomaceous earth or talc with water and trowel the poultice over the stain. If the stain remains when the poultice is removed after 2 or 3 days, repeat the process as necessary.

(b) Oil. If the oil is freshly spilled, soak it up with absorbent paper; do not wipe it up. Cover the stain with a dry powdered material such as portland cement, hydrated lime, cornmeal, or cat litter. Wait approximately 24 hr, then sweep it up. Scrub the remaining stain with scouring powder or a strong soap solution. If the stain is old, cover it with flannel soaked in a solution of equal parts acetone and amyl acetate. Cover the flannel with a pane of glass or a thin concrete slab for 10 to 15 min. Repeat if necessary. Rinse when the cleaning process is complete.

(c) Grease. Scrape the grease from the surface. Scrub with scouring powder, strong soap or detergent, or sodium orthophosphate. If the stain persists, make a stiff poultice with one of the chlorinated solvents. Repeat if necessary. Rinse.

(d) Dirt. Most dirt can be removed with plain water or with a soft brush and water containing a mild soap. If a stronger solution is necessary, use 19 parts water to 1 part hydrochloric acid. If the dirt contains a lot of oil, use the methods for removing lubricating oil. Also, steam cleaning is generally effective for removing dirt. If the dirt is clay, scrape off all that has hardened. Scrub the stain with hot water containing sodium orthophosphate.

(e) Mildew. Mix powdered detergent and sodium orthophosphate with commercial sodium hypochlorite solution and water. After applying the mixture, wait a few days and then scrub the area. Rinse with clear water. Caution: sodium hypochlorite solution bleaches colored clothing and may corrode metal.

(f) Asphalt. Chill molten asphalt with ice (in summer). Scrape or chip it off while it is brittle. Then scrub the area with abrasive powder and rinse thoroughly with water. Do not apply solvents to emulsified asphalt as they will carry the emulsions deeper into the concrete. Scrub with scouring powder and rinse with water. Use a poultice of diatomaceous earth or talc and a solvent to remove cutback asphalt. When the poultice has dried, brush it off. Repeat if necessary.

(g) Efflorescence. Most efflorescence can be removed soon after it forms by washing or by a scrub brush and water. After the efflorescence has begun to build up a deposit, it can be removed by light water blasting or light sandblasting and hosing with clean water. However, some salts become water insoluble shortly after reaching the atmosphere. Efflorescence from these salts can be removed with a dilute solution of hydrochloric or phosphoric acid. Since an acid solution may slightly change the appearance of concrete or masonry, entire walls should be treated to avoid blotching. Only a 1-to 2-percent solution should be used on integrally colored concrete; stronger solutions may etch the surface, revealing the aggregate and hence changing color and texture.

(h) Soot. Scrub the stain with water and scouring powder, powdered pumice, or grit. If this treatment does not remove the stain, swab the area with trichloroethylene and apply a bandage made of three or four layers of cotton material soaked in trichloroethylene. If the stain is on a horizontal surface, hold the bandage against the stain with concrete slabs or stones. If the surface is vertical, prop the bandage against the stain. Periodically, remove, wring out, resaturate, and replace the bandage. Several treatments may be needed. Note: trichloroethylene is highly toxic and can react with fresh concrete, or other strong alkalis, to form dangerous gases. An alternative to the bandage is a poultice. Mix sodium hypochlorite (commercial household bleach is about 5 percent hypochlorite) or diluted Javelle water with talc or other fine material to make a paste. Spread the paste on the stain and allow it to dry thoroughly. Brush off the residue. Repeat the treatment if necessary. Note: sodium hypochlorite and Javelle water will bleach colored clothing and are corrosive to metals.

(i) Graffiti. Apply a proprietary cleaner that contains an alkali, a solvent, and detergent. After scrubbing the graffiti with a brush, leave the cleaner in place for the time indicated by the manufacturer. Rinse thoroughly. Avoid contact with skin. A less expensive, nonproprietary cleaner is dichloromethane, which can be washed off with water. The procedure is the same as with a proprietary cleaner.

c. Environmental considerations. In addition to the potentially adverse worker health and safety effects, improper handling and disposal of cleaning materials and their associated solvents may have adverse environmental effects. Reasonable caution should guide the use of cleaning activities involving the use of potentially hazardous and toxic chemical substances (REMR Technical Note EI-M-1.2 (USAEWES 1985h)). Manufacturer's directions and recommendations for the protection of environmental quality should be carefully followed. The MSDS should be consulted for detailed handling and disposal instructions. The MSDS also provides guidance on appropriate responses in the event of spills. In cases where the effects of a chemical substance on environmental quality are unknown, chemical substances should be treated as potentially hazardous or toxic materials. Residual cleaning solutions may be classified as a hazardous waste, requiring special disposal considerations. The MSDS will generally recommend that Federal, state, and local regulations be consulted prior to determining disposal requirements. Improper handling and disposal of waste materials may result in civil and criminal liability.

7-3. Surface Coatings and Sealing Compounds

Surface coatings and sealing compounds are applied to concrete for protection against chemical attack of surfaces by acids, alkalies, salt solutions, or a wide variety of organic chemicals. Coatings and sealers may also be used to reduce the amount of water penetration into concrete and as a decorative system for concrete. Thick filled coatings are occasionally used to protect concrete from physical damage. Before a protective coating or sealer is used on concrete, it should be determined that the concrete actually needs protection. The cause and extent of the deterioration, the rate of attack, the condition of the concrete, and the environmental factors must all be considered in the selection of a coating or sealer. For example, application of an impermeable coating or overlay may, under certain conditions, trap moisture within the concrete, thereby doing more harm than good (Section 6-16). Information on the susceptibility of concrete to chemical attack and selection, installation, and inspection of surface barrier systems is provided by ACI 515.1R, Pinney (1991), Bean (1988), and Husbands and Causey (1990).

a. Surface preparation. Proper concrete surface preparation is the single most important step for successful application of a coating. The concrete surface must be sound, clean, and dry before the coating is applied. Surface contaminants such as oils, dirt, curing compounds, and efflorescence must be removed. After the contaminants are removed, any unsound surface concrete must be removed before the concrete is coated.

(1) The most common method for determining the soundness of a concrete surface is the pipe-cap pulloff test (ACI 503R). Other commercial pulloff equipment such as the DYNA tester is satisfactory. Oils and other deep surface contaminants may have to be removed by chemical or steam cleaning. Abrasive blasting, shotblasting, high-pressure water, mechanical scarifiers, and acid cleaning are the methods must often used to remove the unsound surface concrete as well as most contaminants. Acid etching should be used only when other methods of surface preparation are impractical.

(2) Materials used to repair substrate surface defects should be compatible with the coating to be used. A latex-modified mortar should not be used if the coating to be used is solvent based. If epoxy resins are used, they should be highly filled and the surfaces should be slightly abraded before the coating is applied. Most coatings require a dry surface. Poor adhesion of a coating can result if water vapor diffuses out to the concrete surfaces.

(3) Some ASTM Standard Practices and Test Methods which may be helpful in preparing and inspecting concrete surfaces for coatings are listed below:

- Standard Practice for Surface Cleaning Concrete for Coating, ASTM D 4258.

- Standard Practice for Abrading Concrete, ASTM D 4259.

- Standard Practice for Acid Etching Concrete, ASTM D 4260.

- Standard Practice for Surface Cleaning Concrete Unit Masonry for Coating, ASTM D 4261.

- Test Method for pH of Chemically Cleaned or Etched Concrete Surfaces, ASTM D 4262.

- Test Method for Indicating Moisture in Concrete by the Plastic Sheet Method, ASTM D 4263.

b. Coatings. Factors to be considered in selection of a coating include intended function of the coating, properties of the coating, application conditions, anticipated service conditions, and life cycle costs. Coating properties that may be important, depending on the specific application, include abrasion resistance, water or chemical resistance, flexibility, curing time, temperature range, and aesthetics. ACI 515.1R and NACE International Standard RP0591-91 (NACE 1991) provide information on generic types of coatings that are appropriate for various exposure conditions. Candidate coating systems should be thoroughly evaluated to ensure that they are appropriate for the intended function and meet other desired characteristics such as ease of application and aesthetics. A test patch applied to the intended substrate in an area where the coating will be subjected to anticipated service conditions is recommended.

(1) General considerations.

(a) Typically, coating thicknesses range from a few mils to 3 mm (125 mils) or more, depending on the purpose of the coating. Thin coatings (<1 mm (40 mils)) are normally used for dampproofing, mild chemical attack, and for decorative coatings. Thick coatings (>1mm (40 mils)) are used for waterproofing, as protection against severe chemical attack, and as protection from physical damage.

(b) Coatings with very low permeabilities may do more harm than good by increasing the level of moisture in concrete if water enters the concrete from the side not coated (Section 6-16). Some coatings do transmit water vapor (breath) and these should be selected if it is expected that water will enter from the uncoated side of the concrete.

(c) Most coatings will not bridge cracks in concrete, but there are some elastomer coatings (polyurethanes and acrylics) that will bridge narrow cracks (<0.8 mm (<1/32-in)). Some thin polymer coatings (high-molecular-weight methacrylates and a few epoxy resins) are formulated to seal cracks in horizontal concrete structures by gravity.

(2) Characteristics of coatings. Characteristics of selected coatings for concrete that prevent attack from corrosive chemicals in the atmosphere and reduce moisture penetration are discussed in the following and summarized in Table 7-1 (NACE 1991).

(a) Silicones, siloxanes, and silanes are best used as water repellents. These materials are not designed to resist chemical attack or physical abuse.

(b) Cementitious coatings may be decorative products and are usually modified with latex for use in mild chemical exposure conditions. Certain inorganic silicate cements may be used to waterproof concrete from the positive or negative side.

(c) Thin film urethanes (up to 0.13 mm (5 mils) per coat) are used to seal concrete for nondusting, cleanability, graffiti resistance, and resistance to mild chemicals. They are used for dry interior exposures on walls and floors that have moderate physical abuse and for exterior weathering. Urethanes are available in two forms: aliphatic urethanes for color and gloss retention in exterior sunlight exposure and aromatic urethanes for exposures other than sunlight and UV light, or where ambering and chalking are acceptable.

(d) Epoxy polyesters are thin film coatings (up to 0.08 mm (3 mils) per coat) designed for color, nondusting, cleanability, and resistance to water for a brief period. They are used primarily for interior and exterior exposures on walls that experience little physical abuse.

(e) Latexes are coatings used for color, appearance, and cleanability. For exterior use, acrylic latexes provide improved color and gloss retention (vinyl latexes are not normally recommended because they tend to hydrolyze under high pH situations). Elastomeric formulations (e.g., acrylic, silicone), which provide waterproofing and crack bridging properties, are also available.

(f) Chlorinated rubbers are thin film coatings designed for color, nondusting, cleanability, and resistance to water and mild chemicals; chlorinated rubbers may chalk on exterior weathering exposures, unless modified.

Table 7-1
Guidance on Selection of Concrete Coatings to Prevent Chemical Attack and Reduce Moisture Penetration (NACE 1991)

Coating	Water Repellancy	Cleanability	Aesthetic	Concrete Dusting	Mild Chemical	Severe Chemical	Moderate Physical	Severe Physical
Silicones/Silanes/Siloxane	R	NR	NR	NR	NR	NR	NR	NR
Cementitious	R	NR	R	NR	NR	NR	NR	NR
Thin-Film Polyurethane	R	R	R	R	R	NR	R	NR
Epoxy Polyester	R	R	R	R	R	NR	NR	NR
Latex[1]	R	R	R	R	NR[2]	NR	NR[2]	NR
Chlorinated Rubber	R	R	R	R	R	NR	R	NR
Epoxy	R	R	R	R	R	R	R	NR
Epoxy Phenolic	R	R	R	R	R	R	R	R
Aggregate Filled Epoxy	R	R	R	R	R	R	R	R
Urethane Elastomers	R	R	R	R	R	R	R	R
Epoxy or Urethane Coal Tar	R	R	NR	R	R	R	R	R
Vinyl Ester/Polyester	R	R	NR	R	R	R	R	R

Note: Reprinted with permission from NACE International. The complete edition of NACE Standard RP0591-91 is available from NACE International, P. O. Box 218340, Houston, Texas 77218-8340, phone: 713/492-0535, fax: 713/492-8254.

R = Recommended
NR = Not Recommended

---
[1] Excluding vinyl latices
[2] Certain latices may be suitable for service

NOTE: The recommendations provided are general. Candidate coating systems must be thoroughly evaluated to ensure that they are appropriate for the intended service conditions and meet other desired characteristics. The above list is not necessarily all-inclusive.

(g) Epoxies are two component products that are available in thin film (less than 0.25 mm (10 mils)) and thick film (0.25 to 1.27 mm (10 to 50 mils)) coatings. Epoxies have excellent adhesion to dry concrete, and epoxies have the ability to seal porous concrete and bug holes. Epoxies also exhibit good chemical resistance, hardness, and abrasion resistance. Epoxies are typically used for interior chemical and physical abuse conditions, because they tend to chalk and fade in atmospheric and sunlight exposure. Epoxy formulations that develop good adhesion to wet surfaces are also available.

(h) Epoxy phenolics are two component products similar to epoxies. They are phenolic modified to improve their chemical resistance. They are normally used for severe chemical environments and as floor coatings.

(i) Aggregate-filled epoxies are thick film coatings (3.18 mm (125 mils) or more thickness) that are usually applied by spray, trowel, or aggregate broadcast methods. Normally used in areas of severe physical abuse, these epoxies are still resistant to mild and severe chemicals. They are excellent floor coatings for areas of severe physical abuse. Floor toppings can be made aesthetically pleasing through selection of the appropriate color and type of aggregate.

(j) Thick film elastomers (up to 3.18 mm (125 mils)), such as urethane (ASTM D 16 Type V) and polysulfide, are normally applied by spray, trowel, or self-leveling methods. Normally used in areas of severe physical abuse that require a flexible coating, the rubber-like film displays excellent resistance to impact damage and the ability to bridge hairline cracks in concrete.

(k) Epoxy- or urethane-coal tars are moderately thick coatings (0.38 to 0.76 mm (15 to 30 mils)) with excellent water and good chemical resistance that are normally applied with a sprayer. The black color may restrict their usage for aesthetic reasons.

(l) Vinylesters and polyesters are moderately thick coatings (0.76 to 1.27 mm (30 to 50 mils)) with excellent resistance to acids and strong oxidizers that are applied by spray or trowel. Thicker films may be obtained with silica floor fillers and reinforcing fabric or mat.

(m) Coatings, such as inorganic silicate cementitious products, sulphur concrete, polysulfide elastomers, epoxy polysulfides, and others, also offer protection to concrete exposed to atmospheric and aggressive environments such as secondary containment structures.

(3) Application.

(a) The manufacturer's recommended application rate and method of application should be followed when a coating is applied to concrete. The surface profile and porosity will have an effect on the application rate. A test patch is useful in determining the surface preparation, application rate, and appearance of a particular concrete coating.

(b) The temperature of the concrete should be constant or dropping when some coatings are applied to avoid blisters or pin holes caused by the expansion of gases inside of the concrete. The temperature of the concrete should be above the dewpoint while the coating is curing to prevent water condensation on the coating.

c. Sealers. Sealers are thin, nonfilled liquids that penetrate or form a thin film (less than 0.13 mm (5 mils)) on concrete. Sealers are used for water repellency when there is no hydrostatic pressure, for dust control, and for reducing the amount of water soluble salts that enter into concrete. Penetrating sealers, such as the silanes and siloxanes, are recommended for areas subjected to traffic. Some sealers do not change the appearance of the concrete, but others may darken the surface. Some sealers are slightly volatile, and high winds and temperatures during application may affect their performance. Concrete sealers that have not been approved for the type of concrete masonry units (CMU) in service should not be used on CMU. Liquid surface treatments known as hardeners should be used only as emergency measures for treatment of deficiencies in hardened concrete floor slabs. They are not intended to provide additional wear resistance in new, well designed, well constructed, and cured floors (ACI 302.1R).

Pfeifer and Scali (1981) have provided what is probably the most comprehensive report on the sealer properties that are relevant for bridge concrete. The basic findings of this study were confirmed in subsequent work by Kottke (1987) and Husbands and Causey (1990). A performance-based specification for concrete sealers on bridges was developed by Carter (1993).

7-4. Joint Maintenance

Little maintenance is required for buried sealants such as waterstops because they are not exposed to weathering and other deteriorating influences. Most field-molded sealants will, however, require periodic maintenance if an effective seal is to be maintained and deterioration of the structure is to be avoided. The necessity for joint main-tenance is determined by service conditions and by the type of material used.

Minor touchups of small gaps and soft or hard spots in field-molded sealants can usually be made with the same sealant. However, where the failure is extensive, it is usually necessary to remove the sealant and replace it. A sealant that has generally failed but has not come out of the sealing groove should be removed by hand tools or, on large projects, by routing or plowing with suitable tools. To improve the shape factor, the sealant reservoir may be enlarged by sawing. After proper preparation has been made to ensure clean joint faces and additional measures designed to improve sealant performance, such as improvement of shape factor, provision of backup material, and possible selection of a better type of sealant, have been accomplished, the joint may be resealed. For additional information on joint sealant materials, joint design, and installation of sealants, see EM 1110-2-2102, and ACI 504R.

Chapter 8
Specialized Repairs

8-1. Lock Wall Rehabilitation

Approximately one-half of the Corps' navigation lock chambers were built prior to 1940. Consequently, the concrete in these structures does not contain intentionally entrained air and is therefore susceptible to deterioration by freezing and thawing. Since more than 75 percent of these older structures are located in the U.S. Army Engineer Divisions, North Central and Ohio River, areas of relatively severe exposure to freezing and thawing, it is not surprising that many of these structures exhibit significant concrete deterioration. Depending upon exposure conditions, depths of concrete deterioration can range from surface scaling to several feet. The general approach in lock wall rehabilitation has been to remove 0.3 to 0.9 m (1 to 3 ft) of concrete from the face of the lock wall and replace it with new portland-cement concrete using conventional concrete forming and placing techniques (McDonald 1987b). Other approaches that have been used include shotcrete, preplaced-aggregate concrete, and precast concrete stay-in-place forms. Also, a variety of thin overlays have been applied to lock walls.

a. Cast-in-place concrete. The economics of conventional cast-in-place concrete replacement compared to other rehabilitation techniques usually depends on the thickness of the concrete section to be replaced. For sections in the range of 152 to 305 mm (6 to 12 in.), both formed and nonformed techniques such as shotcrete are economically competitive. When the thickness of the replacement section exceeds 305 mm (12 in.), conventional formwork and concrete replacement are generally more economical. Conventional cast-in-place concrete has several advantages over other rehabilitation materials. It can be proportioned to simulate the existing concrete substrate, thus minimizing strains due to material incompatibility; proper air entrainment in the replacement concrete can be obtained by use of admixtures to ensure resistance to cycles of freezing and thawing; and materials, equipment, and personnel with experience in conventional concrete application are readily available in most areas.

(1) Surface preparation. Removal of deteriorated concrete is usually accomplished by drilling a line of small-diameter holes along the top of the lock wall parallel to the removal face, loading the holes with light charges of explosive (usually detonating cord), cushioning the charges by stemming the holes, and detonating the

explosive with electric blasting caps. Blasting plans (holes spacing, size of charges, necessity for delays in detonation of charges, etc.) are developed based on results of previous work or test blasts. After the existing concrete is removed, the lock walls are sounded to locate any areas of loose or deteriorated concrete extending beyond the removal line. Such concrete is removed by chipping, grinding, or water blasting. The bonding surface must be clean and free of materials that could inhibit bond.

(2) Concrete anchors. Dowels are normally used to anchor the new concrete facing to the existing concrete walls and to position vertical and horizontal reinforcing steel in the concrete facing. Design criteria for use in design of dowels for anchoring relatively thin sections (less than 0.8 m (2.5 ft)) of cast-in-place concrete facing were developed based on laboratory and field tests (Liu and Holland 1981).

(a) Dowels should be No. 6 deformed reinforcing bars conforming to ASTM A 615. Typical dowel spacing is 1.2 m (4 ft) center to center in both directions, except that a maximum dowel spacing of 0.6 m (2 ft) center to center may be specified in the vicinity of local openings and recesses, and along the perimeters of monoliths. The band width of this dowelling should be at least 0.6 m (2 ft).

(b) If the average compressive strength of three drilled cores obtained from the existing concrete as determined in accordance with ASTM C 42 (CRD-C 27) is less than 21 MPa (3,000 psi), the embedment length should be determined by conducting field pullout tests. For existing concrete with an average compressive strength equal to or greater than 21 MPa (3,000 psi), dowel embedment length should be a minimum of 15 times the nominal diameter of the dowel unless a shorter embeddment length can be justified through field pullout tests. Dowels should be embedded in holes drilled with rotary-percussive equipment. The drill holes must be clean and free of materials that can inhibit bond. Either a nonshrink grout conforming to ASTM C 1107 (CRD-C 621) or an epoxy resin-based bonding system conforming to ASTM C 881 (CRD-C 595), Type I, or other approved bonding systems should be used as the bonding agent. The embeddment length for dowels in the new concrete facing should be determined in accordance with ACI 318.

(c) A minimum of 3 dowels per 1,000 to be installed should be field tested with the testing dispersed over the entire surface area to receive dowels. Test procedures should be similar to those described by Liu and Holland

(1981). The embedment length is considered adequate when the applied pullout load is equal to or greater than the calculated yield strength for the dowel in all tests.

(d) Prepackaged polyester-resin grout has been used to embed the anchors on most projects, and field pullout tests on anchors installed under <u>dry</u> conditions indicate this to be a satisfactory procedure.

(3) Reinforcement. Mats of reinforcing steel, usually No. 5 or 6 bars on 305-mm (12-in.) centers each way, are hung vertically on the dowels (Figure 8-1). Concrete cover over the reinforcing is usually 100 or 127 mm (4 or 5 in.). In some cases, the reinforcing mat, wall armor, and other lock wall appurtenances are installed on the form prior to positioning the form on the face of the lock wall (Figure 8-2).

(4) Concrete placement. Once the reinforcement and formwork are in position, replacement concrete is placed by pumping or discharging it directly into hoppers fitted with various lengths of flexible pipe commonly known as elephant trunks. Lift heights varying from 1.5 m (5 ft) to full face of approximately 15 m (50 ft) have been used. Normally, concrete is placed on alternating monoliths along a lock chamber wall. Generally, forms are removed 1 to 3 days following concrete placement and a membrane curing compound applied to formed concrete surfaces.

(5) Performance. One of the most persistent problems in lock wall rehabilitation resulting from use of this

a. Wall armor and concrete reinforcement in place on the form

b. Concrete forming being placed in position on the lock wall

Figure 8-2. Lock wall rehabilitation, Brandon Rock Lock

Figure 8-1. Placing reinforcing steel and horizontal armor prior to concrete placement, Locks and Dam No. 3, Monogahela River

approach is cracking in the replacement concrete. These cracks, which generally extend completely through the conventional replacement concrete, are attributed primarily to the restraint provided through bond of the new concrete to the stable mass of existing concrete. As the relatively thin layer of resurfacing concrete attempts to contract as a result of shrinkage, thermal gradients, and autogenous volume changes, tensile strains develop in the replacement concrete. When these strains exceed the ultimate tensile strain capacity of the replacement concrete, cracks develop. The lock monolith on the left in Figure 8-3 was resurfaced with a 457-mm (18-in.) overlay of new concrete. As a result of the restraint provided by the existing concrete, extensive cracking developed. The monolith on the right was completely removed and reconstructed in 1.5-m (5-ft) lifts with the same concrete materials and mixture proportions. Without an existing concrete substrate to provide restraint, no cracking occurred.

(a) A finite element analysis of a typical lock wall resurfacing (Norman, Campbell, and Garner 1988) also demonstrated the effect of restraint by the existing concrete. Under normal conditions of good bond between the two concretes, stress levels sufficient to cause cracking in the replacement concrete developed within about 3 days (Figure 8-4a). In comparison, a bond breaker at the interface resulted in almost no stress in the replacement concrete. The analysis also indicated that shrinkage was a significant factor (Figure 8-4b). Even the lower bound shrinkage resulted in tensile stresses in excess of 2.1 MPa (300 psi).

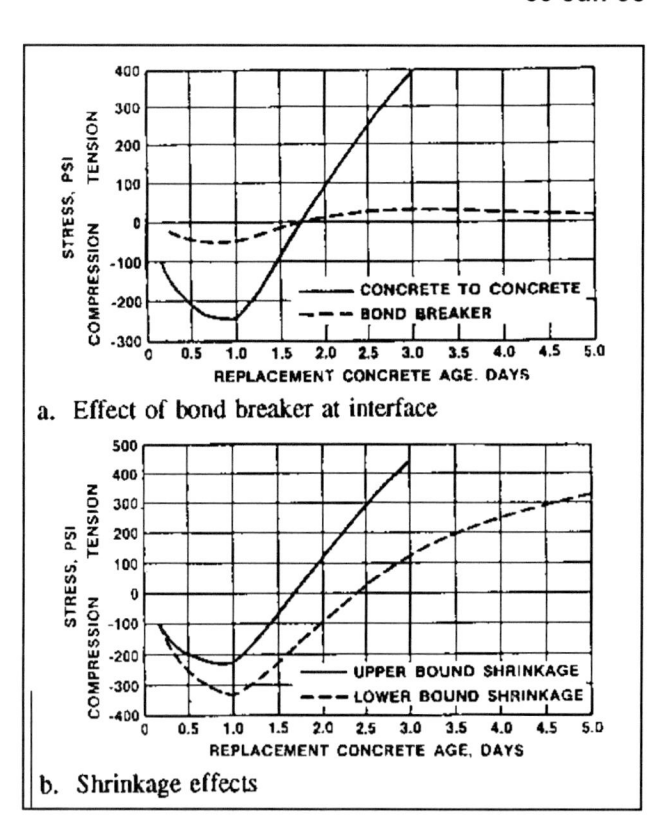

a. Effect of bond breaker at interface

b. Shrinkage effects

Figure 8-4. Results of a finite element analysis of a typical lock wall resurfacing

(b) CEWES-SC recommendations to minimize shrinkage and install a bond breaker (Hammons, Garner, and Smith 1989) were implemented by the U.S. Army Engineer District, Pittsburgh, during the rehabilitation of Dashields Locks, Ohio River. An examination of the project indicated that cracking of concrete placed during 1989 was significantly less than that of concrete placed during the previous construction season prior to implementation of CEWES-SC recommendations. Also, resurfacing of the lock walls at Lock and Dam 20, Mississippi River, resulted in significantly less cracking in the conventional concrete than was previously experienced at other rehabilitation projects within the U.S. Army Engineer District, Rock Island. The reduced cracking was attributed to a combination of factors including lower cement content, larger maximum size coarse aggregate, lower placing and curing temperatures, smaller volumes of placement, and close attention to curing (Wickersham 1987). Preformed horizontal contraction joints 1.5 m (5 ft) on center were effective in controlling horizontal cracking in the rehabilitation of Lock No. 1, Mississippi River (McDonald 1987b).

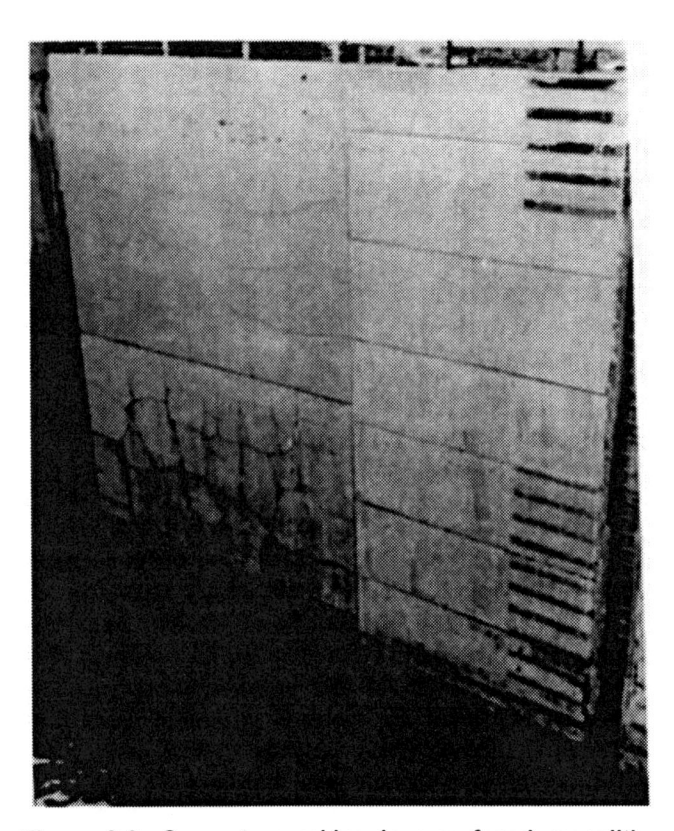

Figure 8-3. Concrete cracking in resurfaced monolith (left) compared to reconstructed monolith, Lock No. 1, Mississippi River

b.   Shotcrete.

(1)   General considerations. For repair of sections less than 152 mm (6 in.) thick, shotcrete is generally more economical than conventional concrete because of the saving in forming costs. Properly applied shotcrete is a structurally adequate and durable material, and it is capable of excellent bond with concrete and other construction materials. These favorable properties make shotcrete an appropriate selection for repair in many cases. However, there are some concerns about the use of shotcrete to rehabilitate old lock walls. The resistance of shotcrete to cycles of freezing and thawing is generally good despite a lack of entrained air. This resistance is attributed in part to the low permeability of properly proportioned and applied shotcrete which minimizes the ingress of moisture thus preventing the shotcrete from becoming critically saturated. Consequently, if the existing nonair-entrained concrete in a lock wall behind a shotcrete repair never becomes critically saturated by migration from beneath or behind the lock wall, it is likely that such a repair will be successful. However, if moisture does migrate through the lock wall and the shotcrete is unable to permit the passage of water through it to the exposed surface, it is likely that the existing concrete will be more fully saturated during future cycles of freezing and thawing. If frost penetration exceeds the thickness of the shotcrete section under these conditions, freeze-thaw deterioration of the existing nonair-entrained concrete should be expected.

(2)   Performance.

(a)   The river chamber at Emsworth Locks and Dam, opened to traffic in 1921, was refaced with shotcrete in 1959. By 1981 the shotcrete repair had deteriorated to the stage shown in Figure 8-5. Spalling appeared to have originated in the upper portion of the wall where the shotcrete overlay was relatively thin and surface preparation minimal. Deterioration apparently propagated down the wall to a point where the shotcrete was of sufficient thickness (approximately 100 mm (4 in.)) to contain dowels and wire mesh. Horizontal cores from the chamber walls showed the remaining shotcrete to be in generally good condition (Figure 8-6a). However, the original concrete immediately behind the shotcrete exhibited significant deterioration, probably caused by freezing and thawing. Cores of similar concrete from the land chamber which did not receive a shotcrete overlay were in generally good condition from the surface inward (Figure 8-6b). This appears to be an example of an overlay contributing to the saturation of the original concrete with increased deterioration resulting from freezing and thawing.

a. Overall view

b. Closeup view

Figure 8-5. Spalling of shotcrete used to reface river chamber wall, Emsworth Locks and Dam

(b)   Anchored and reinforced shotcrete was used to reface wall areas of both the Davis and Sabin Locks at Sault Ste. Marie, Michigan. After 30 to 35 years in service, the shotcrete appeared to be in generally good condition. However, cores taken from the lock walls in 1983 showed deterioration of the nonair-entrained concrete immediately behind the shotcrete in several cases (Figure 8-7a). In comparison, concrete in wall areas which were not refaced was in good condition (Figure 8-7b). Laboratory tests of selected cores indicated the permeability of the shotcrete to be approximately one-third that of the concrete. Also, the shotcrete contained

a. Portions of the core from two drill holes through river chamber wall which had been refaced with shotcrete

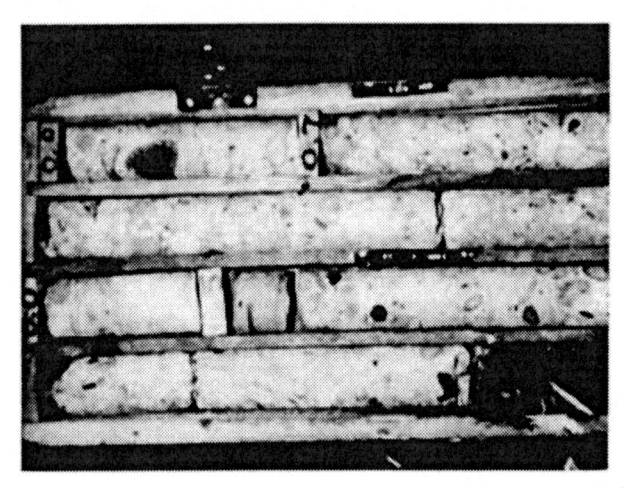

b. Portions of the core from two drill holes through land chamber wall which was not refaced

Figure 8-6. Comparison of cores taken horizontally from lock chamber walls, Emsworth Locks

1.8-percent air voids compared to 0.9 percent in the concrete. In areas where moisture migration did not cause the concrete behind the shotcrete to become critically saturated during cycles of freezing and thawing, the shotcrete was well bonded to the concrete and there was no evidence of deterioration in either material. However, when cores from such areas were subjected to accelerated freeze-thaw testing in a saturated state, the concrete completely disintegrated while the shotcrete remained generally intact (Figure 8-8). Similar cores subjected to accelerated freeze-thaw testing in a dry state remained substantially intact (Figure 8-9).

(c) Portions of the chamber walls at Dresden Island Lock were repaired in the early 1950's with anchored and good condition after about 40 years in service. This good performance relative to other shotcrete repairs is attributed primarily to the thickness of the shotcrete overlay which apparently exceeded the depth of frost penetration. Three horizontal cores taken through the repaired walls in the mid-1970's indicated that the overlay is a minimum of 305 mm (12 in.) thick. The shotcrete-concrete interface was intact and the concrete immediately behind the shotcrete exhibited no deterioration. After more than 40 years in service, the average depth of deterioration in the unrepaired concrete walls as determined by petrographic examination was approximately 215 mm (8-1/2 in.). Assuming that the depth of frost penetration is approximately equal to the depth of deterioration, the thickness of shotcrete is about 50 percent greater than the depth of frost penetration. Therefore, the concrete behind the shotcrete would not be expected to exhibit freeze-thaw deterioration even though it may have been critically saturated. Air-void data determined according to ASTM C 457 (CRD-C 42) indicated the shotcrete had about 3 percent total air with approximately 2 percent of it in voids small enough to be classified as useful for frost resistance. The air-void spacing factors ranged from 0.25 to 0.36 mm (0.010 to 0.014 in.). While these values are larger than is desirable (0.20 mm (0.008 in.) is considered the maximum value for air-entrained concrete), they may have imparted some frost resistance. Also, there were no large voids or strings of voids resulting from lack of consolidation such as have been observed with other shotcrete specimens.

(d) The concrete in the walls of Lower Monumental Lock has an inadequate air-void system to resist damage caused by freezing and thawing while critically saturated. After approximately 10 years in service, the concrete had deteriorated to the point that the aggregate was exposed. After laboratory tests and a full-scale field test which demonstrated that a 10-mm (3/8-in.)-thick latex-modified fiberglass-reinforced spray-up coating had the potential to meet repair requirements, which included time constraints on the lock downtime that precluded conventional concrete replacement (Schrader 1981), the lock walls were repaired (Figure 8-10) in 1980. In March 1983 several small isolated areas of debonded coating (less than 1 percent of the total area) were identified in the lock. The coating in these locations was removed and the areas resprayed similar to the original repair. Also, a fairly large debonded zone in monolith 11 was repaired. A September 1983 inspection reported the original coating to

a. Concrete deterioration immediately behind shotcrete overlay

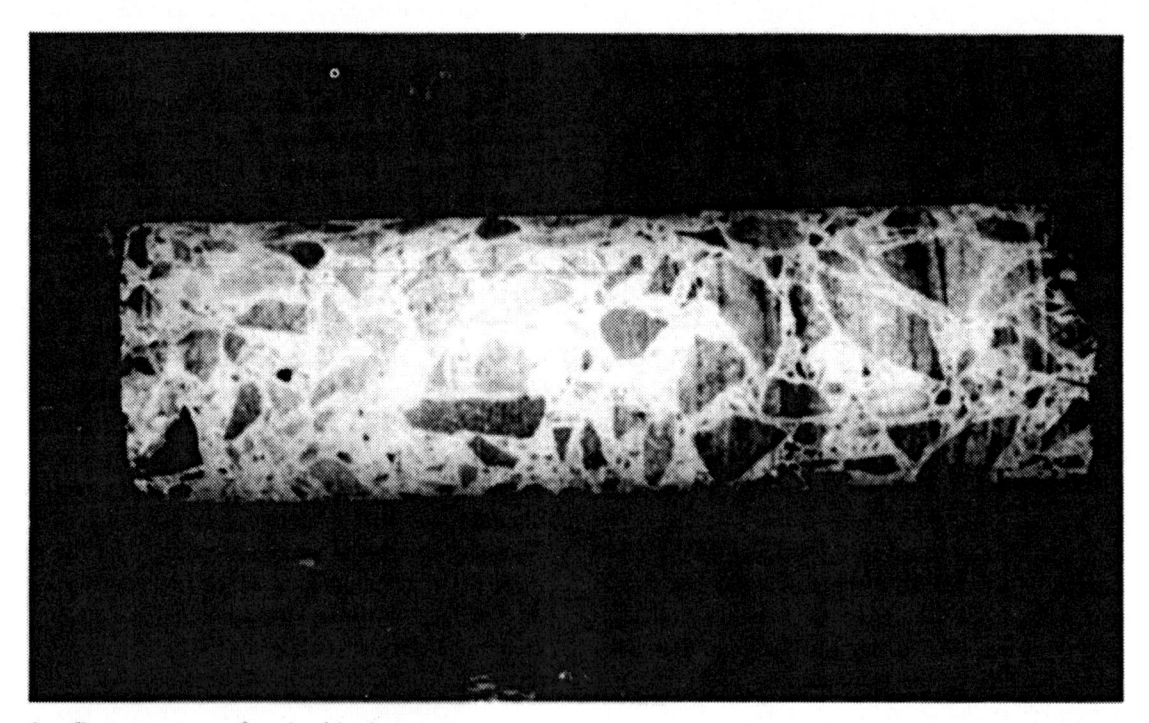

b. Concrete not refaced with shotcrete

Figure 8-7. Comparison of cores taken from lock chamber walls, Davis and Sabin Locks

a. Before freezing and thawing

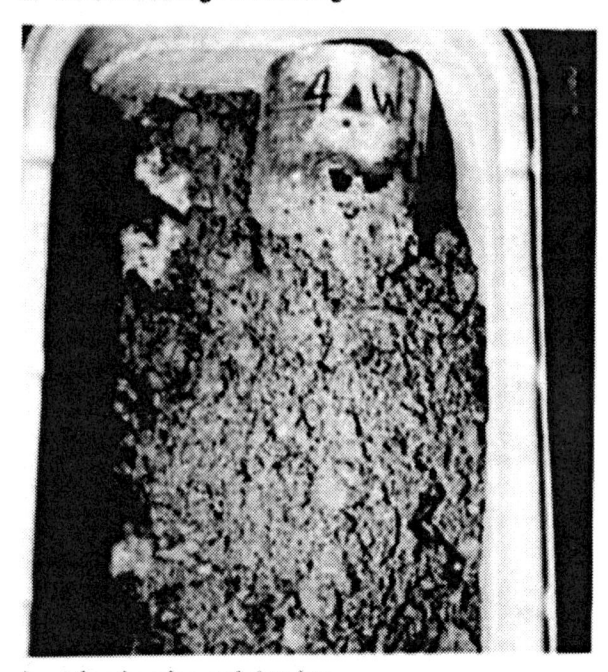

b. After freezing and thawing

Figure 8-8. Combination shotcrete and concrete core before and after 35 cycles of freezing and thawing in a saturated condition

a. Before freezing and thawing

b. After freezing and thawing

Figure 8-9. Combination shotcrete and concrete core before and after 35 cycles of freezing and thawing in a dry condition

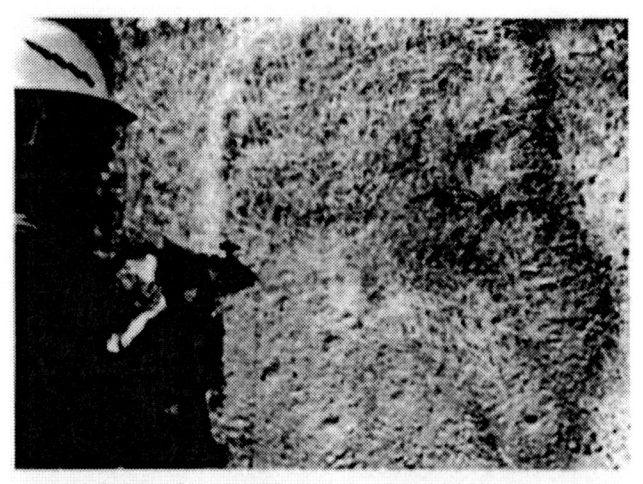

a. Applying the coating with equipment that simultaneously sprays the mortar from one nozzle while it chops and blows the fiber from a cutterhead

b. Typical appearance of the lock wall before and after repair

Figure 8-10. Application of coating to lock wall, Lower Monumental Lock

be in generally good condition. However, the coatings applied earlier in 1983 had all failed after only 6 months in service. The major difference in the two repairs was the type of latex used. Saran was used in the original work, whereas styrene butadiene was used in the repairs which failed. It was also reported that delaminations were common where latex spray-up material had been applied over hardened latex from a previous coating. One possibility for this failure is that the high-pressure water jet did not properly clean the surface of latex film prior to application of the subsequent layer. After approximately 5 years in service, large pieces of the spray-up coating began to fall from the lock walls. Failure occurred in the

concrete substrate immediately behind the bond line. Moisture migrating toward the surface from within the mass concrete was trapped at the interface because of the low water vapor transmission characteristics of the coating. Subsequent ice and hydraulic pressure caused the coating to debond (Schrader 1992).

c. *Preplaced-aggregate concrete.* Since drying shrinkage and creep occur almost exclusively in the cement paste fraction of concrete and since both phenomena are resisted by the aggregate, particularly if the coarse aggregate particles are in point-to-point contact, drying shrinkage and creep are both remarkably less for preplaced-aggregate concrete than for conventionally placed concrete. The reduction in drying shrinkage reduces the probability of cracking under conditions of restrained shrinkage, but it is at the expense of a reduced capability to relax concentrated stresses through creep. The dimensional stability of preplaced-aggregate concrete makes it attractive as a material for the rehabilitation of lock walls and appurtenant structures, particularly if it is successful in mitigating or eliminating the unsightly cracking commonly experienced with conventionally placed concrete. Its potential is further enhanced by the fact that it can be conveniently formed and placed under-water and that it can be grouted in one continuous operation so that there are no cold joints. Laboratory investigation has shown preplaced-aggregate concrete to be superior to conventionally placed, properly air-entrained concrete inresistance to freeze-thaw damage and impermeability (Davis 1960). Cores drilled through preplaced-aggregate concrete overlays have shown excellent bond with the parent concrete.

(1) Applications.

(a) Preplaced-aggregate concrete has been used extensively and effectively to construct or rehabilitate many different kinds of structures, but the technique is not one that many contractors have had experience with. Successful execution of preplaced-aggregate work requires a substantial complement of specialized equipment mobilized and operated by a seasoned crew working under expert supervision. This coupled with the fact that preplaced-aggregate concrete requires stronger and tighter formwork generally results in higher bid prices, as much as one-third higher than conventional concrete according to current bid prices. Therefore, if preplaced-aggregate concrete is the desired repair material, it must be specified uniquely and not as an alternate.

(b) Constructed during the period 1907-1910, the lock chamber walls at Lock No. 5, Monongahela River, required refacing in 1950. The plans called for removal of approximately 450 mm (18 in.) of old concrete from an area extending from the top of the lock walls to about 450 mm (18 in.) below normal pool elevation and the refacing of this area with reinforced concrete. Specifications required that the concrete have a minimum 28-day compressive strength of 24 MPa (3,500 psi) and provided that concrete could be placed by either conventional or preplaced aggregate methods. Also, it was required that one of the two lock chambers be open to navigation at all times. The low bid ($85,000) was submitted by Intrusion-Prepakt Co. to whom the contract was awarded. The low bid was almost $60,000 lower than the second lowest bid. The contractor elected to perform the work without constructing cofferdams. This method necessitated the removal of existing concrete below pool elevation "in the wet," and required that the new concrete below pool elevation be placed behind watertight bulkheads to exclude pool water from the spaces to be filled with concrete. The contractor elected to use the preplaced-aggregate method for concrete placement. Details of the repair operation were reported by Minnotte (1952).

(c) Preplaced-aggregate concrete was used in 1987 to resurface the lock chamber walls at Peoria Lock (Mech 1989). Following removal of a minimum of 305 mm (12 in.) of concrete, the walls were cleaned with high-pressure water, anchors and reinforcing steel were installed, and forms were positioned on individual monoliths. Typical resurfaced areas were about 3 m (10 ft) high and 12 m (40 ft) wide. The forms were filled with water to reduce breakage of the coarse aggregate during placement. The forms were vibrated externally with handheld equipment during grout intrusion which took about 8 to 10 hr of pumping for each monolith. Forms on two of the approximately 30 monoliths had to be reset because anchors holding the forms to the lock wall failed.

(2) Performance.

(a) During the Lock No. 5 repair, preplaced-aggregate concrete test cylinders were made by filling steel molds with coarse aggregate and pumping the mortar mixture into the aggregate through an insert in the base of the molds. Compressive strengths of these 152- by 305-mm (6- by 12-in.) specimens ranged from 20 to 30 MPa (2,880 to 4,300 psi) with an average strength of 26 MPa (3,800 psi). The average compressive strength of cores, drilled from the refacing concrete 1 year later, was 37 MPa (5,385 psi).

(b) Lock No. 5 was removed from service and, with the exception of the land wall, razed in conjunction with the construction of Maxwell Lock and Dam in 1964. A visual examination of the remaining wall in July 1985 showed that the preplaced-aggregate concrete had some cracking and leaching (Figure 8-11) but overall appeared to be in generally good condition after 35 years exposure. This repair demonstrates that the preplaced-aggregate method of concrete placement is a practical alternative to conventional methods for refacing lock walls.

a. Overall view

b. Closeup view

Figure 8-11. Condition of lock chamber wall 35 years after refacing with preplaced-aggregate concrete

(c) Although the repair at Peoria Lock was generally successful, the preplaced-aggregate concrete exhibited some cracking. Some of the cracks could be traced to areas where the contractor had problems with the soaker system used for moist curing. High temperatures during this period ranged between 35 and 38 °C (95 and 100 °F). Some bug holes on the concrete surface were attributed to the method used to vibrate the forms and changing the grout flow requirement from 18 sec to 26 sec.

(d) The bid price for 466 cu m (610 cu yd) of preplaced-aggregate concrete at Peoria Lock was $1,250 per cu m ($960 per cu yd, which included form work, aggregate, placement, grouting operation, and finishing). In comparison, typical bid prices for conventional concrete used on other rehabilitation projects in the U.S. Army Engineer District, Rock Island, have been in the range of $590 to $850 per cu m ($450 to $650 per cu yd) and higher for difficult construction situations.

d. Thin overlays. A variety of thin overlays have been applied to lock walls (McDonald 1987b). In most cases, they have been used in areas where the depth of deterioration was minimal, and apparently the intent was to protect the existing concrete or to improve the appearance of the structure. There has been little, if any, concrete removal associated with these applications. Thin overlays have had very little success with a number of failures during the first winter after application. Such overlays are particularly susceptible to damage by barge impact and abrasion.

e. Precast concrete. Compared with cast-in-place concrete, precasting offers a number of advantages including minimal cracking, durability, rapid construction, reduced future maintenance costs, and improved appearance. Also, precasting minimizes the impact of adverse weather and makes it possible to inspect the finished product prior to its incorporation into the structure. Precast-concrete stay-in-place forms have been used successfully in a number of lock wall rehabilitation projects. In addition to significant reductions in the length of time a lock must be closed to traffic for rehabilitation, the precast-concrete stay-in-place forming systems has the potential to eliminate the need for dewatering a lock chamber during wall resurfacing (ABAM Engineers 1989). Additional information on applications of precast concrete in repair and replacement of a variety of civil works structures is given in Section 8-5.

(1) Development. A precast concrete stay-in-place form system for lock wall rehabilitation was developed as part of the REMR Research Program. The objectives of this work were to develop a precast concrete rehabilitation system which provides superior durability with minimal cracking, accommodates all of the normal lock hardware and appurtenances, minimizes lock downtime, and can be implemented at a wide variety of project sites. To accomplish these goals, the system was required to satisfy a well-defined set of durability, functional, constructibilty, and cost/schedule criteria (McDonald 1987a).

(a) A wide range of alternatives for achieving the design objectives was evaluated through a process of value engineering (ABAM Engineers 1987a). Based on this analysis, it was concluded that the most advantageous combination of design alternatives was a precast-quality concrete (minimum compressive strength of 45 MPa (6,500 psi)), conventionally reinforced, flat panel, horizontally oriented and tied to the lock wall (Figure 8-12). A typical panel was 165 mm (6-1/2 in.) thick, 1.8 m (6 ft) wide, 9.1 m (30 ft) long, and it weighed approximately 7 tons. The panels are tied to the lock monolith along the top and bottom edges with form ties designed to support the loads of the infill concrete placement.

(b) To demonstrate the constructibility of the system, eight panels of varying sizes were precast and installed on two one-half-scale simulated lock monoliths (Figure 8-13). The purpose of the demonstration was to evaluate the feasibility of the precast concrete stay-in-place forming system without the risk and investment of undertaking a full-scale lock rehabilitation. Results of this work demonstrated that the precast concrete stay-in-place forming system is a viable method for lock wall rehabilitation (ABAM Engineers 1987b). In addition to providing a concrete surface of superior durability with minimal cracking, the construction cost was very competitive with the cost of conventional cast-in-place concrete. The demonstration also identified a number of areas where the design and installation procedures could be enhanced thus reducing both the cost and schedule associated with the stay-in-place forming system. Development of the precast concrete stay-in-place forming system is summarized in a video report (McDonald 1988) which is available on loan from the CEWES Library.

(2) Applications. To date, the precast concrete stay-in-place forming system has been used to rehabilitate four lock chambers, and contracts for two additional projects have been awarded. In addition, precast concrete panels have been used to overlay the back side of the river wall of two locks. Detailed descriptions of these applications (McDonald and Curtis in preparation) are summarized in the following.

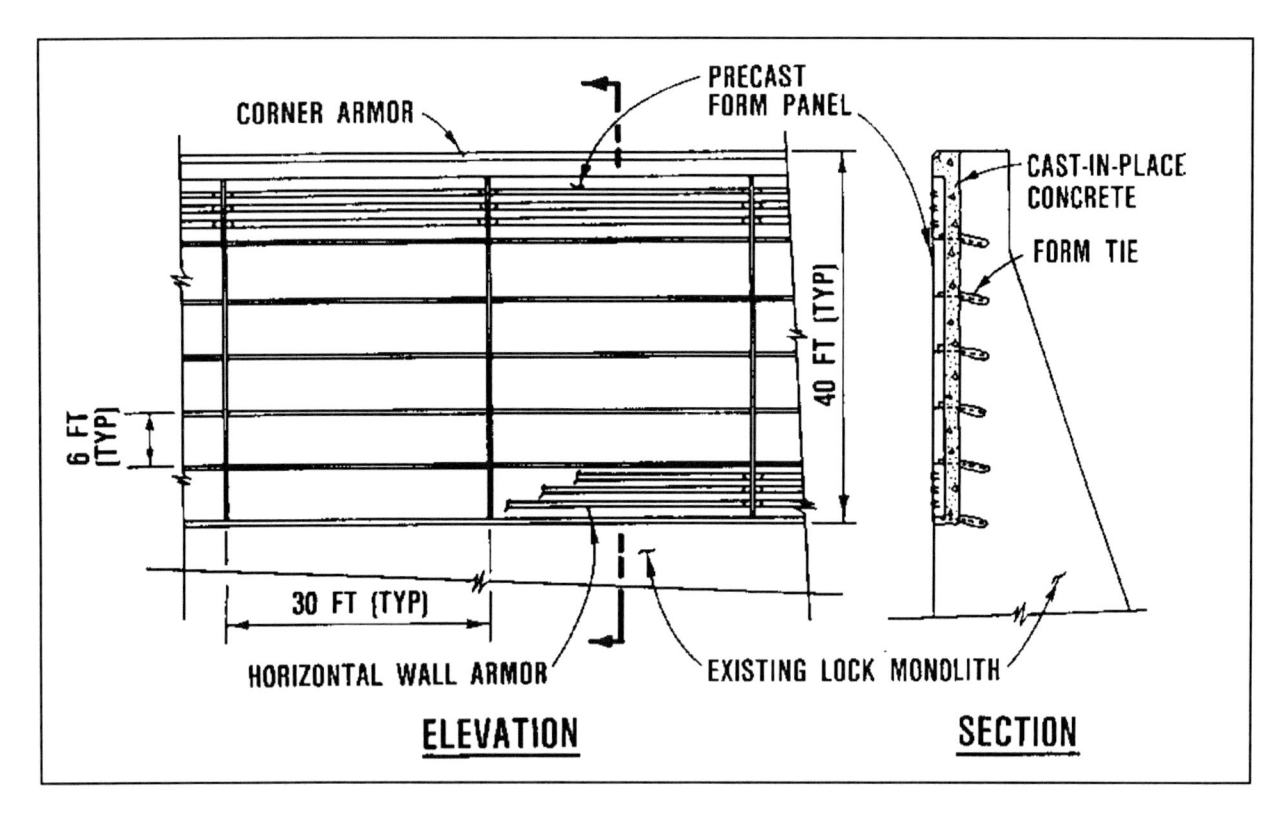

Figure 8-12. Precast concrete stay-in-place forming system for lock wall rehabilitation

Figure 8-13. Demonstrating the constructibility of the precast concrete stay-in-place forming system

(a) The initial application of the precast-concrete stay-in-place forming system for lock chamber resurfacing was at Lock 22, Mississippi River, near Hannibal, MO, during Jan and Feb 1989. A total of 41 panels were precast in a Davenport, IA, plant with a concrete mixture with an average compressive strength of 54 MPa (7,800 psi). The reinforced-concrete panels were 165 mm (6-1/2 in.) thick, 3.2 m (10-1/2 ft) high, and varied in length from 5.8 to 10.7 m (19 to 35 ft). Weld plates at the top and bottom and leveling inserts along the bottom were embedded in each panel during precasting.

(b) Concrete removal began with a horizontal saw cut at the bottom of the repairs. Line drilling, which was accomplished prior to lock closure, and blasting techniques were used to remove a minimum of 216 mm (8-1/2 in.) of concrete from the face of the lock walls. Following final cleanup of the exposed concrete surface, anchors of weldable-grade reinforcing steel were installed to coincide with weld plates in the bottom of the panels.

(c) The panels were transported to the site by river barges and lifted into position with a barge-mounted crane (Figure 8-14). The panels were held on the sawcut ledge while they were leveled with the vertical alignment hardware and the bottom panel anchors were welded to the weld plates embedded in the panels. The top panel anchors were then installed and welded to the embedded weld plates. Each panel was aligned and welded in 2 to 3 hr. The space between the panels and lock wall was filled with nonshrink cementitious grout. Approximately 1,070 sq m (11,500 sq ft) of the lock walls were resurfaced. The bid price for the precast panel repair, including concrete removal, was $980 per sq m ($91 per sq ft). Bid prices for this initial application of the precast concrete stay-in-place forming system for lock wall rehabilitation were somewhat higher than anticipated. However, the rehabilitation went very smoothly, despite severe winter weather conditions, and the precast panels were installed in about one-half the time that would have been required for cast-in-place concrete.

(d) This application of the precast concrete stay-in-place forming system demonstrated the potential for eliminating the need to close and dewater a lock during rehabilitation. Consequently, concepts for installation of the system in an operational lock were developed as part of the REMR research program (ABAM 1989). A mobile cofferdam was selected as the preferred installation method and a final design completed for this concept. This study indicated that it is feasible to repair the walls in an operational lock with only minimal impact on costs.

Figure 8-14. Installation of precast concrete stay-in-place forms at Lock 22, Mississippi River

lock wall repairs, they are potentially applicable to other walls requiring underwater repair.

(e) After the lock was reopened to navigation, the concrete along vertical monolith joints began to exhibit cracking and spalling caused by barge impact and abrasion. Recessing the joints by cutting the concrete with a diamond saw eliminated this problem. The cuts were started at a 5-deg angle to the chamber face and 305 mm (12 in.) outside of the joints to produce a 25-mm (1-in.)-deep recess at the joints. Cracked concrete along the joints was removed with a chipping hammer, and the spalled areas were repaired with a latex-modified cement dry pack.

(f) Based on the experience gained at Lock 22, a number of revisions were incorporated into the design of the precast panel system used at Troy Lock, Hudson River, during the winter of 1991-92. These revisions included new lifting, alignment, anchorage, and joint details (Miles 1993). Also, the design was based on an allowable crack width of 0.15 mm (0.006 in.) compared to 0.25 mm (0.01 in.) at Lock 22.

(g) A total of 112 precast concrete panels were required to resurface the lock chamber walls. A typical panel was 190 mm (7-1/2 in.) thick, 3.6 m (11 ft 10 in.) high, and 6.1 m (20 ft) long. Ten special panels were required to accommodate line poles and ladders. These panels were 521 mm (1 ft 8-1/2 in.) thick except at the recesses where the thickness was reduced to 190 mm (7-1/2 in.). The panels were fabricated with a 25 by 305 mm (1 by 12 in.) taper and a 25 mm (1 in.) chamfer along the vertical joints to reduce impact spalling. Reinforcing mats with 51 mm (2 in.) of concrete cover were placed at both faces of each panel. Panel anchors were hoop-shaped reinforcing bars which extended 102 mm (4 in.) from the rear face of the panel. Two erection anchors were embedded in the top of each panel. A concrete mixture proportioned with a 0.31 water-cementitious ratio for a compressive strength of 48 MPa (7,000 psi) at 28 days was used to precast the panels. The panels were fabricated in a heated building during the winter of 1991-92.

(h) A minimum of 305 mm (12 in.) of concrete was removed from the lock walls by saw cutting around the perimeter of the removal area, line drilling, and explosive blasting. Following air-jet cleaning of the surface, anchors were grouted into holes drilled into the existing concrete. These bent reinforcing bar anchors were located to line up with the hoop anchors embedded in the precast panels. The majority of the panels were installed during the winter of 1991-92 when ambient temperatures were generally cold, at times reaching -18 °C (0 °F). The precast panels were lifted from delivery trucks with a crane and lowered onto steel shims placed on the sawcut ledge. The panels were positioned with temporary holding and adjustment brackets, shebolt form anchors at the top and bottom of the panels at 1.2-m (4-ft) spacings, and steel strongbacks spanning between the shebolts. Once the panels were positioned and aligned, vertical reinforcing bars were manually installed from the top down to intersect the panel and wall anchors (Figure 8-15). Once the bars were properly located, they were tied to the anchors to prevent misalignment during placement of the infill concrete.

Figure 8-15. Installation of precast concrete stay-in-place forms at Troy Lock, Hudson River (Miles 1993)

(i) The top of the precast panels was capped with 0.6 m (2 ft) of reinforced concrete, cast-in-place. A 152-mm (6-in.)-thick apron slab was also placed on top of the existing monolith wall to reduce water penetration and improve durability. Joints between panels and concrete cap and between cap and apron slab were sealed with joint sealant. Upon completion of the lock chamber resurfacing, an inspection of the precast panels revealed that 11 of 112 panels exhibited some fine cracks. Those cracks (four locations) with widths in excess of 0.15 mm (0.006 in.) were repaired by epoxy injection. In contrast, the miter gate monoliths previously resurfaced with cast-in-place concrete exhibited extensive cracking (Figure 8-16).

Figure 8-16. Typical cracking in miter gate monoliths resurfaced with cast-in-place concrete, Troy Lock, Hudson River (Miles 1993)

Figure 8-17. Installing precast concrete panels used to overlay the back side of the river wall at Troy Lock, Hudson River (Miles 1993)

(j) The contractor's bid price for the precast concrete resurfacing in the lock chamber was only $355 per sq m ($33 per sq ft) at Troy Lock compared to $980 per sq m ($91 per sq ft) at Lock 22. Also, the mean bid price for precast concrete at Troy Lock was approximately $50 per sq m ($5 per sq ft) lower than the mean bid price forcast-in-place concrete during the same period. Although the contractor at Troy was inexperienced in both lock rehabilitation and the use of precast concrete, the project progressed quite smoothly and the efficiencies of using precast concrete became very obvious as the work was completed. It is anticipated that as the number of qualified precasters increases and as contractors become more familiar with the advantages of precast concrete, the costs of the precast concrete stay-in-place forming system will be reduced.

(k) In addition to resurfacing the lock chamber, precast concrete panels were used to overlay the back side of the river wall at Troy Lock (Figure 8-17). Original plans for repair of this area required extensive removal of deteriorated concrete and replacement with shotcrete. This repair, which would have had to be accomplished in the dry, would have required construction of an expensive cofferdam to dewater the area. Therefore, it was decided that concrete removal could be minimized and the need for a cofferdam eliminated if this area was repaired with precast concrete panels. Three rows of precast panels were used in the overlay. The panels were installed in 1992 while the lock was in operation. The bottom row of panels was partially submerged during installation and

infill concrete placement. An antiwashout admixture allowed the infill concrete to be effectively placed underwater without a tremie seal having to be maintained. The application of precast concrete in this repair resulted in an estimated savings of approximately $500,000 compared to the original repair method. Also, the durability of the aesthetically pleasing precast concrete should be far superior to shotcrete which has a generally poor performance record in repair of lock walls.

(l) Precast concrete panels were also used to overlay the lower, battered section of the backside of the river wall at Lockport Lock, Illinois Waterway (Figure 8-18). Before the panels were placed, the concrete surface was cleaned of all loose concrete with a high-pressure water jet. Then the precast concrete base was positioned and anchored along the horizontal surface at the base of the wall. In addition to conventional reinforcement in each direction, the precast panels were prestressed vertically. Embedded items included weld plates at the bottom of the panels and steel angles with embedded bolts at the top of the panels. Typical panels were 152 mm (6 in.) thick, 7.1 m (23.2 ft) high, and 2.3 m (7.5 ft) wide.

(m) The panels were installed with a crane. Steel shims were placed between the panel and the base at the weld plates. At the top of the panel, steel angles aligned with the steel angles embedded in the panel were secured to the wall with bolts. After the bolted connections were tightened at the top of the panel, the steel shims at the bottom of the panel were welded to weld plates in the

a. Installation of precast concrete panels

b. River wall complete with precast concrete panels

Figure 8-18. Backside of river wall, Lockport Lock, Illinois Waterway

base and the panel. The precast headcap sections were lifted into place, set on shims between the top of the panel and the cap, and anchored to the wall. Backer rods and joint sealant were used to seal all joints. The 1,073 sq m (11,550 sq ft) of precast panels, which were installed in the fall of 1989 for a bid price of $226 per

sq m ($21 per sq ft), are currently performing satisfactorily.

f. Summary. In the design of lock wall repairs, the depth of frost penetration in the area should be considered, particularly in repair of old nonair-entrained concrete. It appears that if the thickness of any repair section is less than the depth of frost penetration, freeze-thaw deterioration of the existing nonair-entrained concrete should be expected. This deterioration can drastically affect the performance of repair sections without adequate anchoring systems. Also, the cost of alternative repairs should be carefully evaluated in relation to the desired service life of the rehabilitated structure. Only a few years of good service, at best, should be expected of shotcrete in relatively thin layers. However, the cost of such a repair will be relatively low. In comparison, conventionally formed and placed concrete, shotcrete, and preplaced-aggregate concrete, each properly proportioned and placed in thicknesses greater than the depth of frost penetration, should provide a minimum of 25 years of service but at successively greater initial costs. Precast concrete panels used as stay-in-place forms should provide even greater durability at approximately the same cost as cast-in-place concrete.

8-2. Repair of Waterstop Failures

Nearly every concrete structure has joints that must be sealed to ensure its integrity and serviceability. This is particularly true for monolith joints in hydraulic structures such as concrete dams and navigation locks. Embedded waterstops are used to prevent water passage through the monolith joints of such structures. Traditionally, waterstops have been subdivided into two classes: rigid and flexible. Most rigid waterstops are metallic: steel, copper, and occasionally lead. A variety of materials are suitable for use as flexible waterstops; however, polyvinyl chloride (PVC) is probably the most widely used (EM 1110-2-2102). Waterstops must be capable of accommodating movement parallel to the axis of the waterstop as a result of joint opening and closing caused by thermal expansion and contraction. Also, differing foundation conditions between adjacent monoliths may result in relative movements between monoliths perpendicular to the plane of the waterstop. A review of waterstop failures (McDonald 1986) reported leakage through monolith joints ranging from minor flows to more than 2,270 L/min (600 gal/min). In general, leakage was the result of (1) excessive movement of the joint which ruptures the waterstop, (2) honeycombed concrete areas adjacent to the waterstop, (3) contamination of the waterstop surface which prevents bond to the concrete, (4) puncture of the waterstop or complete omission during construction, and, (5) breaks in the waterstop due to poor or no splices. Since it is usually impossible to replace an embedded waterstop, grouting or installation of secondary waterstops is the remedial measure most often used. Several types of secondary waterstops have been tried with various degrees of success and expense. McDonald (1986) identified more than 80 different materials and techniques that have been used, individually and in various combinations, to repair waterstop failures. Some repairs appear to have been successful, while many have failed. Generally, repairs can be grouped into four basic types: surface plates, caulked joints, drill holes filled with elastic material, and chemically grouted joints. The particular method used depends on a number of factors including joint width and degree of movement, hydraulic pressure and rate of water flow through the joint, environment, type of structure, economics, available construction time, and access to the upstream joint face.

a. Surface plates. A plate-type surface waterstop (Figure 8-19) consists of a rigid plate, generally stainless steel, spanning the joint with a neoprene or deformable rubber backing. The plate is attached with anchors which provide an initial pressure on the deformable pad. Water pressure against the plate provides additional pressure so that the deeper the waterstop is in a reservoir, the tighter the plate presses to the surface of the joint. This type waterstop has had varying degrees of success. At John Day Lock, it has worked relatively well but at Lower Monumental Lock, it has not performed satisfactorily. The major difference between these structures is that at Lower Monumental, the joint movement is more than that at John Day. Potential problems with this type of repair include: (1) loosening of the anchor bolts; (2) reverse hydrostatic pressure from water trapped behind the waterstop when the reservoir drops; (3) mechanical abuse such as a barge tearing off the plate; (4) ice pressure from

moisture trapped behind the plate; and (5) hardening of the flexible pad due to aging (Schrader 1980).

b. Caulked joints. When the reservoir level can be dropped below the elevation to which repairs are necessary, and if the joint opening is not too great, a simple and economical repair may be possible by saw cutting along the joint on the positive pressure, or reservoir side, and then filling the cut with an elastic sealant (Figure 8-20). The sawcut should be wide enough to span the joint and cut about 3 mm (1/8 in.) minimum into the concrete on each side of it. The cut must also be deep enough to penetrate any unsound, cracked, or deteriorated materials. Typically, a cut 13 mm (1/2 in.) wide by 38 mm (1-1/2 in.) deep is acceptable. The cut should follow the joint to its base; otherwise, water migrating underneath the sealant can build up pressure in the joint behind the sealant. To perform satisfactorily, the sealant must set rapidly, bond to cool and damp concrete, and remain flexible under anticipated service conditions, and it must not sag in vertical applications or extrude through the joint under hydrostatic pressure. Once problems of partial extrusion due to inadequate curing were solved, this type of repair exhibited reasonable success for several years at Lower Monumental and Little Goose Locks.

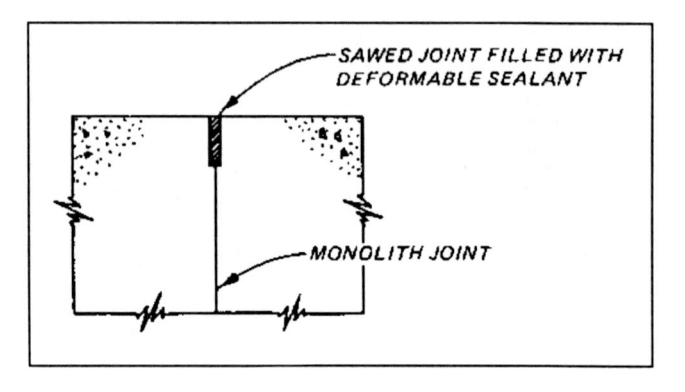

Figure 8-20. Typical sawed joint filled with sealant

c. Drill holes filled with elastic material. This approach (Figure 8-21) consists of drilling a large-diameter hole from the top of the monolith along the joint and into the foundation, and filling the hole with an elastic material. Typically, the hole is 76 to 152 mm (3 to 6 in.) in diameter and drilled by a "down-the-hole" hammer or core drill. The more costly core drilling allows easy visual inspection of hole alignment along the joint, but the down-the-hole hammer has proven to be the preferred method of drilling. Small underwater video cameras can also be used to inspect hole alignment and condition of the joint.

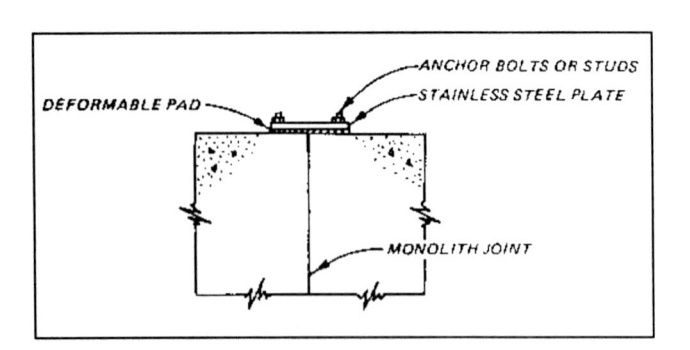

Figure 8-19. Typical plate type waterstop

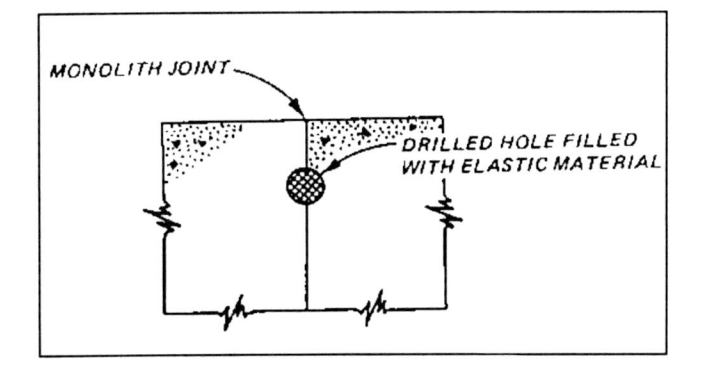

Figure 8-21. Drilled hole with elastic filler

(1) Elastic filler material. The drill holes are filled with an elastic filler material to establish a seal against water penetration through the joint. Criteria for the filler require that it displace water, attain some degree of bond to the concrete surface, remain elastic throughout the life of the structure, be practical for field application, be economical, and have sufficient consistency not to extrude under the hydraulic head to which it will be subjected. If the filler material which is often in liquid form travels out from the drilled hole during placement and into the joint before it sets, better sealing can be expected. The design assumption, however, is that the "poured-in-place" grout filler will form a continuous elastic bulb within the drilled hole. The filler will press tightly to the downstream side of the hole when water pressure is applied to the upstream side, thereby creating a tight seal. Various types of portland-cement and chemical grout have been used as a filler.

(a) Acrylamide grout systems have been used as elastic filler in several applications. Developed primarily for filling voids in permeable sand and gravel foundations, these systems consist of an acrylamide powder, a catalyst, and an initiator. When dissolved in separate water solutions and mixed together, a gelatinous mass results. The reaction time can vary from a few seconds to as much as an hour based on the proportions of the mixture and the temperature. As long as the reacted mass remains in a moist environment, it will stay stable in size and composition and will remain highly flexible. If allowed to dry, its volume can decrease by as much as 90 percent. If resaturated with water, it will regain most of its original properties. The mass can be given more stability, weight, and rigidity through the addition of inert mineral fillers such as diatomaceous earth, bentonite, and pozzolans. In some early applications, portland cement was used to thicken the grout and allow its use in open flowing holes with substantial hydraulic head. However, as the cement hydrated, the grout mass hardened resulting in a

nonflexible filler. Repairs of waterstops using acrylamide grout systems have performed with varying degrees of success. Repairs at Ice Harbor and Lower Monumental Locks, which have been subjected to small relative joint movements, have performed well and are presently functioning as designed. Similar repairs at Little Goose Lock, subject to large relative joint movements, have not performed well. Large movements and high pressures caused the repair material to extrude from the joint in a matter of months. Similar repairs at Pine Flat Dam initially stopped leakage through the repaired joints; however, when the reservoir pool was raised, subjecting the joints to a 61-m (200-ft) head of water, the repairs failed. The potential for extrusion of filler material in unlined drill holes subject to high heads should be recognized.

(b) Water-activated urethane foam grout systems have also been used as elastic fillers. These hydraulic polymers are activated when placed in contact with water and upon curing form a tough, flexible gel approximately 10 times its original volume. One approach in using these materials is to stuff burlap bags which have been saturated with the grout into the drill holes (Figure 8-22). Water in the drill holes activates the grout system causing the grout to expand and completely fill the holes. Tamping the bags as they are pushed into the holes with drill stem will force some of the grout into the joints prior to gel formation. Gel times will vary with ambient and water temperatures but generally can be controlled with additives to be in the range of 5 to 10 min. This procedure was used at Dardanelle Lock with a reduction in gallery leakage of approximately 95 percent. Technical representatives for the material manufacturers should be contacted for detailed guidance on specific grout systems.

(2) Liner materials. In some cases, a continuous tube-type liner has been inserted into the drill hole (Figure 8-23) to contain the filler material. Liner materials include reinforced plastic firehose, natural rubber, elastomer coated fabric, neoprene, synthetic rubber, and felt tubes. In most cases, the liners are not bonded to the walls at the drill holes and rely on differential pressure between the interior of the liner and the external water level to force the liner against the sides of the hole. Obviously, any type of liner material should be of sufficient flexibility to allow the tube to conform to any voids and irregularities in the surface of the drill hole and to accommodate differential movement between adjacent monoliths.

(a) A repair procedure which was developed based on techniques used for in situ relining of pipelines, allows

a. Saturating burlap bags with chemical grout

b. Saturated bag on top of drill hole awaiting insertion

Figure 8-22. Remedial waterstop installation, Dardanelle Lock

bonding of the liner to the surface of the drill hole. In this procedure, the liner is fabricated from thin polyurethane film with an under layer of felt approximately 6 mm (1/4 in.) thick. Prior to installation, water-activated polyurethane resin is poured into the tube to saturate the felt lining. The resin is distributed evenly throughout the tube by passing it through a set of pinch rollers. The tube is then inserted into the drill hole, and as it is inserted water pressure is used to turn the tube inside-out (Figure 8-24). When the inversion process is complete, the resin-impregnated felt is in contact with the surface of

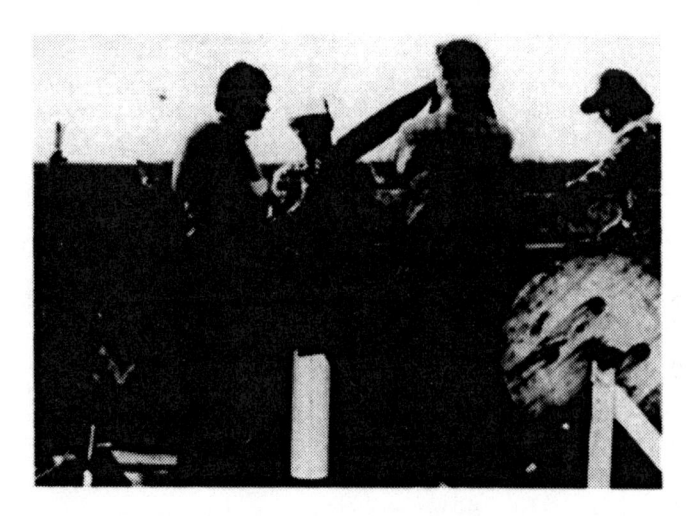

Figure 8-23. Two-component natural rubber and elastomer coated nylon sleeve being inserted into drill hole, Bagnell Dam

the drill hole (Figure 8-25). Water inside the drill hole activates the resin grout creating a bond between the linerand the drill hole. This procedure was used to repair three leaking joints at Pine Flat Dam in March 1985. The drill holes ranged in depth from 58 to 78 m (189 to 257 ft) with a combined flow of 515 L/min (136 gal/min). After repair, the flow dropped to 45 L/min (12 gal/min). As a result of the excellent performance of these repairs, the same procedure was used in 1993 to correct leakage through additional monolith joints at Pine Flat Dam.

(b) A variety of materials have been used as fillers inside the drill hole liners. These include water, bentonite slurry, and various formulations of chemical grout. Criteria for the filler are generally the same as those for drill holes without liners. Filler grouts should have a density greater than that of water and should be placed by tremie tubes starting at the bottom of the liner to displace all water in the liner. Once the filler material is in place, the hole should be capped flush with the top surface. The cap should contain a removable plug for periodic inspection of the filler material.

(c) Video inspections of the drill holes before and after insertion of the liner have proven to be beneficial in determining the exact location of the concrete-foundation interface, irregularities in the concrete surface, seepage locations, and adequacy of the insertion process. Also, if the inversion technique is proposed for liner insertion, the bottom 25 percent of the drill hole should be

Figure 8-24. Inversion apparatus for installing tube in drill hole, Pine Flat Dam

pressure-tested to determine if leakage rates in this zone are sufficient to allow water in the hole to escape during inversion.

  d. Chemically grouted joints. Chemical grouting has been successfully used to seal isolated areas of interior leakage such as around the perimeter of a drainage gallery where it crosses a joint and contraction joints in regulating outlet conduits. Grouting has also been used to seal exterior joints such as those on the upstream face of a dam. Chemically grouted smooth joints often fail when subjected to small relative movement.

  (1) Interior joints. The repair procedure consists of drilling an array of small-diameter holes from various locations within the gallery or conduit to intercept the

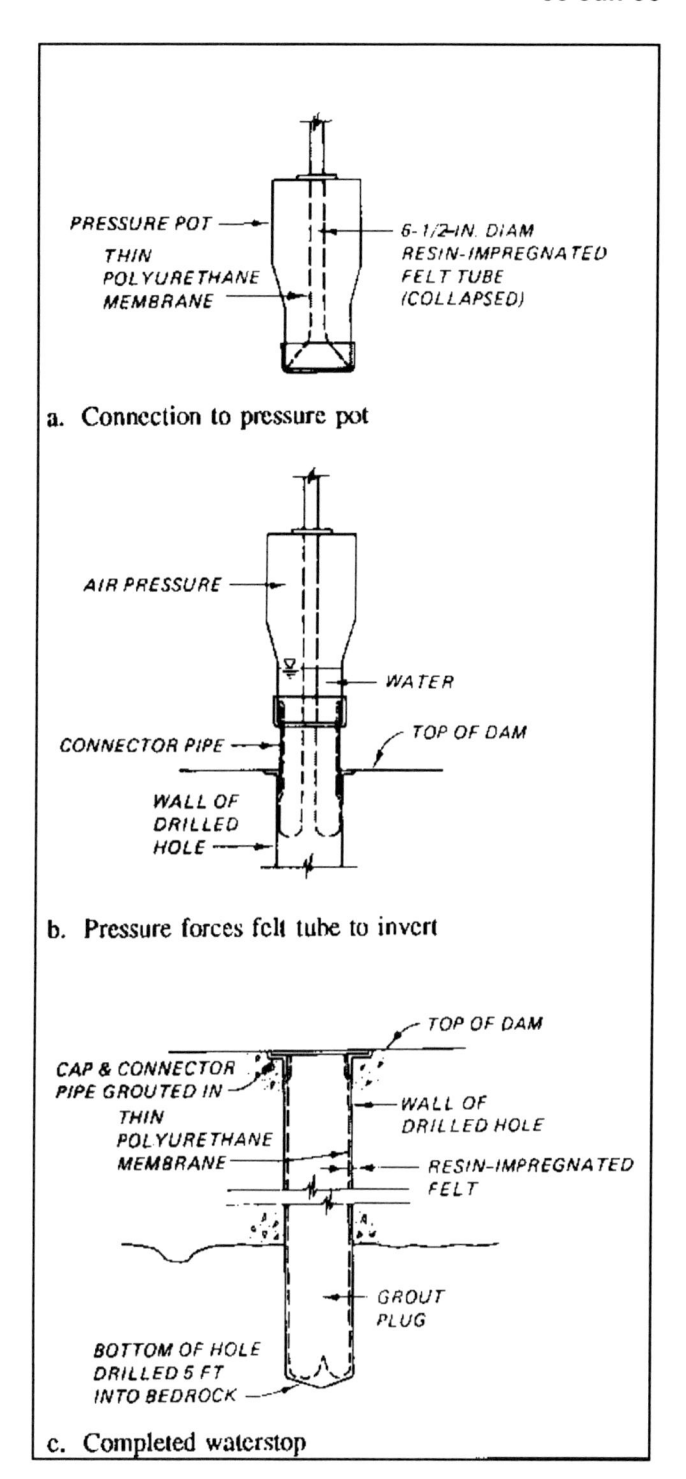

a. Connection to pressure pot

b. Pressure forces felt tube to invert

c. Completed waterstop

Figure 8-25. Remedial waterstop installation, Pine Flat Dam

joint behind the waterstop (Figure 8-26). An elastic chemical grout is then injected through the drill hole. Criteria for the injection grout require that it have low

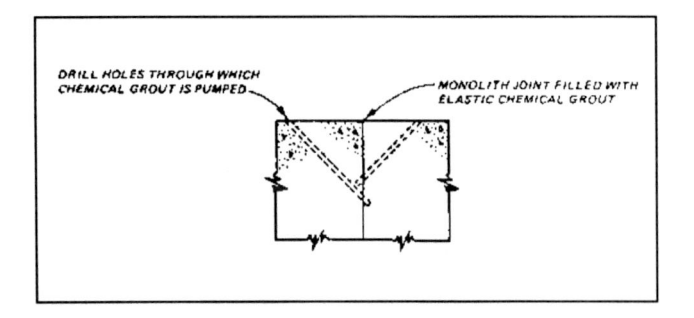

Figure 8-26. Joint filled with elastic grout through injection

viscosity, gel or set quickly, bond to wet surfaces, be suitable for underwater injection, possess good elastic strength, and tolerate unavoidable debris. Water-activated polyurethane injection resins generally meet the desired criteria. Gel times for these materials are normally in the range of 5 to 60 sec which is usually adequate for low-flow conditions. Large-volume or high-pressure flows must be controlled during grout injection and curing. Materials and methods commonly used to control such flows include lead wool hammered into the joint, foam rubber, or strips of other absorbent materials soaked in water-activated polyurethane and packed into the joint, oakum packed into the joint, and small-diameter pipes embedded in the packing material to relieve pressure and divert flow. If the joint opening is greater than approximately 2.5 mm (0.1 in.), a surface plate waterstop may be necessary within the gallery or conduit to prevent grout extrusion with time because of hydraulic pressure along the joint.

(2) Exterior joints. A combination of grouted joint and surface plate waterstop techniques was used to seal vertical contraction joints on the upstream face of the Richard B. Russell Dam (Figure 8-27). A permeable grout tube was placed in the vertical vee along the face of each joint and covered with an elastomeric sealant. After

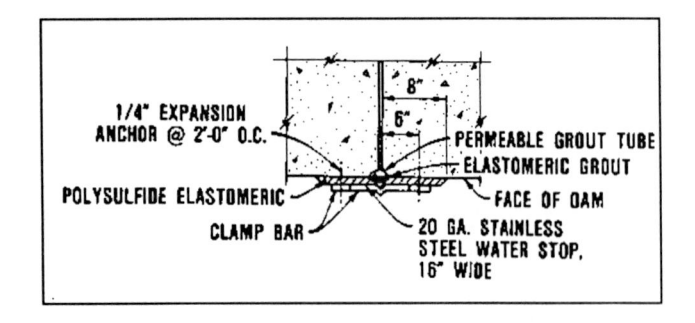

Figure 8-27. Combination grouted joint and surface plate-type waterstop, Richard B. Russell Dam

the sealant hardened, the grout tube was injected with a polysulfide sealant to fill the joints from the dam face into the embedded waterstop. The polysulfide sealant was also placed on the face of the dam for a distance of 203 mm (8 in.) on either side of the joints. Prior to hardening of the sealant, a surface plate waterstop of 20-gauge stainless steel was anchored into position over the joint.

e. Summary. Because of a lack of appropriate test methods and equipment, most of the materials and procedures described have been used in prototype repairs with limited or no laboratory evaluation of their effectiveness in the particular application. Consequently, a test apparatus was designed and constructed, as part of the REMR research program, to allow systematic evaluation of waterstop repair techniques prior to application in prototype structures. The apparatus consists of two concrete blocks, one fixed and one movable, with a simulated monolith joint between the blocks (Figure 8-28). The performance of waterstop repairs can be evaluated for water heads up to 76 m (250 ft) and joint movements up to 10 mm (0.4 in.). Preliminary results of current short-duration tests indicate that grouted joint repairs and most caulked joint repairs begin to leak at joint movements between 1.3 and 2.5 mm (0.05 and 0.1 in.). Catastrophic failure, generally caused by debonding of the repair material, usually occurs at a joint movement of 2.5 mm (0.1 in.) for polyurethane grouts and 5 to 10 mm (0.2 to 0.4 in.) for most joint sealants. To date, the most successful repair consisted of polyurethane grout injection of the joint combined with caulking of the upstream face. This repair exhibited no leakage for joint movements up to 5 mm (0.2 in.) and a water head of 70 m (230 ft).

Figure 8-28. Remedial waterstop test apparatus

Leakage was first observed at a joint movement of 8 mm (0.3 in.) and a water head of 38 m (125 ft); however, the magnitude of the leakage was too small to measure for water heads up to 76 m (250 ft). Additional repair materials and procedures are currently being evaluated and a REMR report will be published upon completion of these tests. In the meantime, available test results can be obtained by contacting U.S. Army Engineer Waterways Experiment Station, Structures Laboratory, Concrete Technology Division, (CEWES-SC).

8-3. Stilling Basin Repairs

A typical stilling basin design includes a downstream end sill from 0.9 to 6 m (3 to 20 ft) high to create a permanent pool for energy dissipation of high-velocity flows. Unfortunately, these pools also serve in many cases to trap rocks, reinforcing steel, and similar debris. In most cases the presence of debris and subsequent erosion damage are the result of one or more of the following: (a) construction diversion flow through constricted portions of the stilling basin; (b) eddy currents created by diversion flows or powerhouse discharges adjacent to the basin; (c) construction activities in the vicinity of the basin, particularly those involving cofferdams; (d) nonsymmetrical discharges into the basin; (e) flow separation and eddy action within the basin to transport riprap from the exit channel into the basin, and (f) topography of the outflow channel (McDonald 1980). While high-quality concrete is capable of resisting high water velocities for many years with little or no damage, the concrete cannot withstand the abrasive grinding action of entrapped debris. In such cases, abrasion-erosion damage ranging in depth from a few inches to several feet can result, depending on the extent of debris and the flow conditions. A variety of repair materials and techniques including armored concrete, conventional concrete, epoxy resins, fiber-reinforced concrete, polymer-impregnated concrete, preplaced-aggregate concrete, silica-fume concrete, and tremie concrete have been used to repair erosion damage in stilling basins. Applications of the various repair materials are described in detail in the 31 case histories reported by McDonald (1980). Selected case histories are summarized in the following.

a. Old River Control Structure. Prefabricated modules of steel plate anchored to the top of the end sill and to the floor slab directly behind the downstream row of baffles were used in repair of the stilling basin at the Old River Control Structure. Thirty modules, 7.3 m (24 ft) long and varying in widths from 0.9 to 6.7 m (3 to 22 ft), were fabricated from 13-m (1/2-in.)-thick steel plate. Vertical diaphragm plates were welded to the horizontal plate, both to stiffen the plate and to provide a formed void in which to retain the grout. Individual modules were positioned and anchored underwater using polyester resin grouted anchors. A portland-cement grout containing steel fibers was then pumped into the modules.

(1) An underwater inspection 8 months after the repairs showed 7 of the 30 modules had lost portions of their steel plate ranging from 20 to 100 percent of the surface area. A number of anchor bolts were found broken either flush with the plate, flush with the grout, or pulled completely out. In those areas where steel plating was lost, the exposed grout surfaces showed no evidence of significant erosion.

(2) A second inspection, approximately 2 years after the repair, revealed that additional steel plating had been ripped from four of the modules previously damaged. Also, an additional nine modules had sustained damage. Minor erosion had occurred in the stilling basin slab upstream from the modules. The stilling basin was reported to be free from rock and other debris. Apparently, any rock or debris discharged through the structure was flushed from the stilling basin over the fillet formed by the modules at the end sill.

(3) Subsequent underwater inspections indicated continuing loss of the steel plate until the stilling basin was dewatered for inspection and repair in 1987, 11 years after the original repair. Following removal of silt which covered most of the stilling basin floor, an inspection indicated that approximately 90 percent of the steel plate was missing. Protruding portions of the remaining plate were cut flush with the surface, and the leading edge of the grout wedge was removed to a minimum depth of 305 mm (12 in.). Isolated areas of erosion damage and spalling in the grout fillet were repaired with shotcrete. The remainder of the basin received a 305-mm (12-in.)-thick overlay of portland-cement concrete (0.45 w/c).

b. Pomona Dam. Prior to repair, a hydraulic model study of the existing stilling basin at Pomona Dam was conducted to investigate discharge conditions which might account for debris in the basin and to evaluate potential modifications to eliminate these conditions. Model tests (Oswalt 1971) confirmed that severe separation of flow from one side wall and eddy action strong enough to circulate stone in the model occurred within the basin for discharges and tailwaters common to the site. Photographs (Figure 8-29) of subsurface upstream flow in the right side of the basin show the results of a discharge rate of 28 cu m/sec (1,000 cu ft/sec). Based on the model

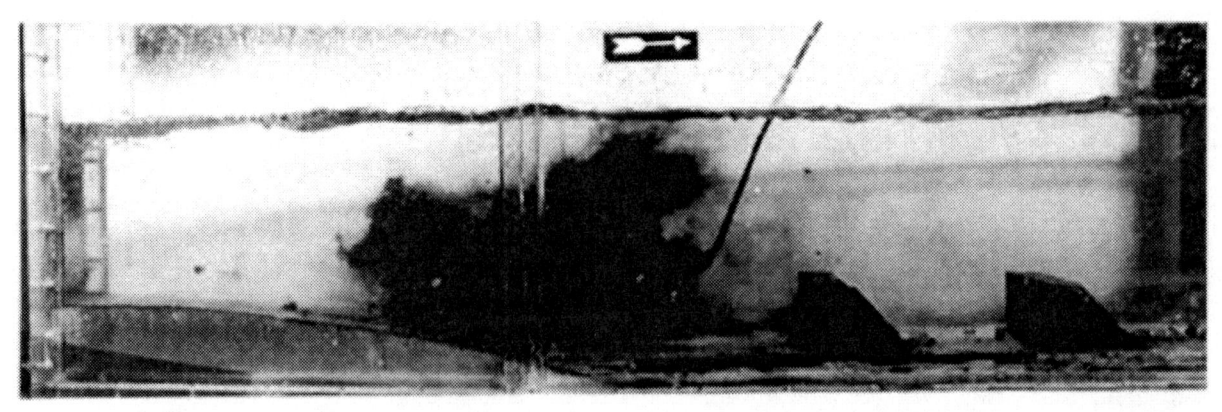

a. Tailwater elevation A

b. Tailwater elevation B

Figure 8-29. Subsurface upstream flow in the stilling basin at Pomona Dam (after Oswalt 1971)

study, it was recommended that the most practical solution was to provide a 0.9-m (3-ft)-thick overlay of the basin slab upstream of the first row of baffles; a 0.5-m (1.5-ft) overlay between the two rows of baffles; and a 1:1 sloped-face to the existing end sill. This solution provided a wearing surface for the area of greatest erosion and a depression at the downstream end of the basin for trapping debris. However, flow separation and eddy action were not eliminated by this modification. Therefore, it was recommended that a fairly large discharge sufficient to create a good hydraulic jump without eddy action be released periodically to flush debris from the basin.

(1) The final design for the repair included (a) a minimum 13-mm (1/2-in.)-thick epoxy mortar applied to approximately one-half of the transition slab; (b) an epoxy mortar applied to the upstream face of the right three upstream baffles; (c) a 0.6-m (2-ft)-thick concrete overlay

slab placed on 70 percent of the upstream basin slab; and (d) a sloped concrete end sill. The reinforced-concrete overlay was recessed into the original transition slab and anchored to the original basin slab. The coarse aggregate used in the repair concrete was Iron Mountain trap rock, an abrasion-resistant aggregate. The average compressive strength of the repair concrete was 47 MPa (6,790 psi) at 28 days.

(2) The stilling basin was dewatered for inspection 5 years after repair. The depression at the downstream end of the overlay slab appeared to have functioned as desired. Most of the debris, approximately 0.8 cu m (1 cu yd) of rocks, was found in the trap adjacent to the overlay slab. The concrete overlay had suffered only minor damage with general erosion averaging about 3 mm (1/8 in.) deep with maximum depths of 13 mm (1/2 in.). The location of the erosion coincided with that occurring

prior to the repair. Apparently, debris was still being circulated at some discharge rate.

(3) The epoxy mortar overlay had not suffered any visible erosion damage; however, cracks were observed in several areas. In one of these areas the epoxy mortar coating was not bonded to the concrete. Upon removal of the mortar in this area (approximately 2 sq m (25 sq ft)), it was observed that the majority of the failure plane occurred in the concrete at depths up to 19 mm (3/4 in.). Following removal, the area was cleaned and backfilled with a low modulus, moisture-insensitive epoxy mortar. In all other areas, even those with cracks, the epoxy mortar overlay appeared to be well bonded.

(4) Based on a comparison of discharge rates and slab erosion, before and after the repair, it was concluded that the repair had definitely reduced the rate of erosion. The debris trap and the abrasion-resistant concrete were considered significant factors in this reduction.

(5) The next inspection, 5 years later, indicated that the stilling basin floor remained in good condition with essentially no damage since the previous inspection. Approximately 4 cu m (5 cu yd) of debris, mostly rocks, was removed from the debris trap at the downstream end of the basin (Figure 8-30).

c. *Nolin Lake Dam.* The conduit and stilling basin at Nolin were dewatered for inspection in 1974. Erosion was reported in the lower portion of the parabolic section, the stilling basin floor, the lower part of the baffles, and along the top of the end sill. The most severe erosion was in the area between the wall baffles and the end sill where holes 0.6 to 0.9 m (2 to 3 ft) deep had been eroded into the stilling basin floor along the sidewalls (Figure 8-31).

(1) Dewatering of the stilling basin for repair was initiated in May 1975. The structural repair work included raising the stilling basin floor elevation 228 mm (9 in.) and raising the end sill elevation 0.3 m (1 ft). Nonreinforced conventional concrete designed for 34-MPa (5,000-psi) compressive strength was used in the repair. A hydraulic model study of the existing basin was not conducted; however, the structure was modified in an attempt to minimize entry of debris into the basin. New work included adding end walls at the end of the stilling basin and a 15-m (50-ft)-long concrete paved channel section (Figure 8-32). Also, a concrete pad was constructed adjacent to the right stilling basin wall to permit a mobile crane to place a closure at the end of the stilling basin wall for more expeditious dewatering of the basin.

(2) A diver inspection of the stilling basin in 1976 indicated approximately 3,600 kg (4 tons) of rock was in the stilling basin. The majority of this rock was from 19 to 127 mm (3/4 to 5 in.) in diameter with some scattered rock up to 305 mm (12 in.) in diameter. The rock, piled up to 0.38 m (1.25 ft) deep, apparently entered the basin from downstream. Also, piles up to 0.46 m (1.5 ft) deep of similar rock were found on the slab downstream from the stilling basin. Erosion up to 203 mm (8 in.) deep was reported for concrete surfaces which were sufficiently cleared of debris to be inspected.

(3) A similar inspection in August 1977 indicated that approximately 900 to 1,350 kg (1 to 1-1/2 tons) of large, limestone rock, all with angular edges, was scattered around in the stilling basin. No small or rounded rock was found. Since the basin had been cleaned during the previous inspection, this rock was first thought to have been thrown into the basin by visitors. When the stilling basin was dewatered for inspection in October 1977, no rock or debris was found inside the basin. Apparently, the large rock discovered in the August inspection had been flushed from the basin during the lake drawdown when the discharge reached a maximum of 208 cu m/sec (7,340 cu ft/sec). No additional damage had occurred in the stilling basin since the 1976 inspection.

(4) Significant erosion damage was reported when the stilling basin was dewatered for inspection in 1984. The most severe erosion was located behind the wall baffles (Figure 8-33) similar to that prior to repair in 1975. Each scour hole contained well-rounded debris. Temporary repairs included removal of debris from the scour holes and filling them with conventional concrete. Also, the half baffles attached to each wall of the stilling basin were removed. A hydraulic model of the stilling basin was constructed to investigate potential modifications to the basin to minimize chances of debris entering the basin with subsequent erosion damage to the concrete.

(5) Results of the model study were incorporated into a permanent repair in 1987. Modifications included rebuilding the ogee section in the shape of a "whales back;" overlaying the basin floor; adding a sloping face to the end sill; raising the basin walls 0.6 m (2 ft); paving an additional 30 m (100 ft) of the retreat channel; slush grouting derrick stone in the retreat channel; and adding new slush grouted riprap beside the basin.

(6) The condition of the concrete was described as good with no significant defects when the basin was dewatered for inspection in August 1988. The maximum

Figure 8-30. Stilling basin dewatered for inspection, Pomona Dam

discharge to that point had been 143 cu m/sec (5,050 cu ft/sec) for a period of 13 days.

d. Kinzua Dam. Because of the proximity of a pumped storage power plant on the left abutment and problems from spray, especially during the winter months, the right-side sluices at Kinzua Dam were used most of the time. This mode of operation caused eddy currents which carried debris into the stilling basin. The fact that

the end sill was below streambed level contributed to the deposition of debris in the basin. As a result, erosion of the concrete to depths of 1 m (3.5 ft) was reported less than 4 years after the basin was placed into normal operation.

(1) The initial repair work (1973-74) was accomplished in two stages; cellular cofferdams enclosed about 60 percent of the stilling basin for each stage, permitting

a. Debris collected behind wall baffle

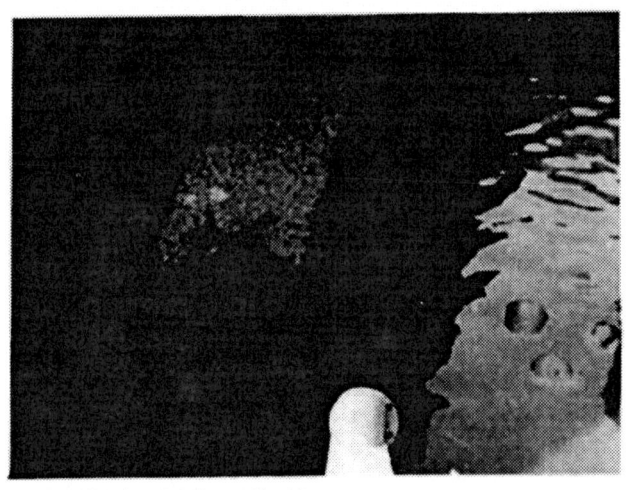

b. Erosion behind wall baffle

Figure 8-31. Concrete erosion damage, Nolin Dam stilling basin, 1974

a. Prior to repair

b. After repair

Figure 8-32. Modifications to outlet work, Nolin Dam, 1975

stream flow in the unobstructed part of the stilling basin. Approximately 1,070 cu m (1,400 cu yd) of fiber-reinforced concrete was required to overlay the basin floor (Figure 8-34). A concrete mixture containing 25-mm (1-in.) steel fibers, proportioned for 8 and 41 MPa (1,100 and 6,000 psi) flexural and compressive strengths, respectively, was used for the anchored and bonded overlay. The overlay was placed to an elevation 0.3 m (1 ft) higher than the original floor from the toe of the dam to a point near the baffles.

(2) The initial diver inspection of the repair in November 1974, 1 year after completion of Stage I repairs, indicated minor concrete deterioration in some areas of the basin floor. An estimated 34 cu m(45 cu yd) of debris was removed from the basin. In an effort to verify the source of this material, approximately 5,500 bricks of three different types were placed in the river downstream of the basin. Six days later, the basin was inspected by divers who found numerous smooth

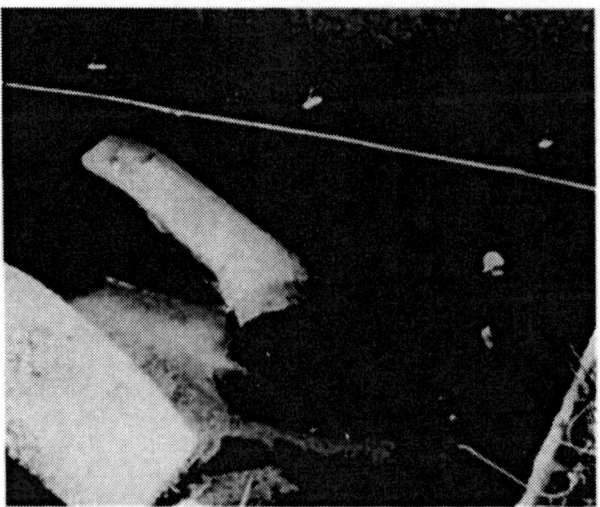

a. Scour holes behind baffles

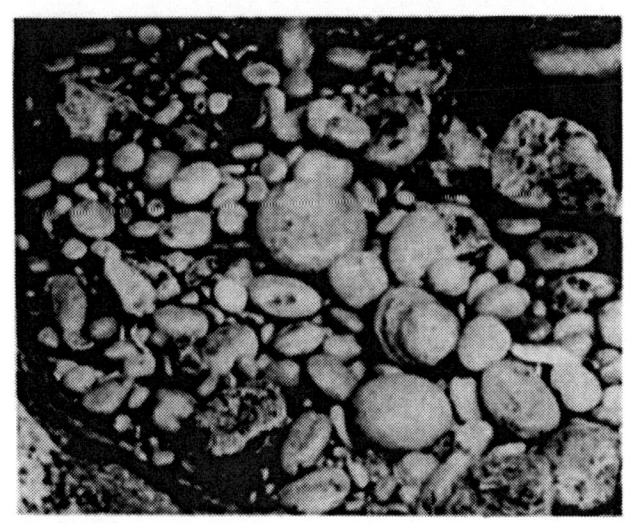

b. Debris in scour holes

Figure 8-33. Concrete erosion damage, Nolin Dam stilling basin, 1984

pieces of all three types of brick in various sizes. This finding proved that the debris was being brought into the stilling basin from areas downstream of the end sill.

(3) In April 1975, additional abrasion-erosion damage to the fiber-reinforced concrete was reported. Maximum depths ranged from 127 to 432 mm (5 to 17 in.). Approximately 34 cu m (45 cu yd) of debris was removed from the stilling basin. Additional erosion was reported in May 1975, and approximately 46 cu m (60 cu yd) of debris was removed from the basin. At this point,

symmetrical operation of the lower sluices was initiated to minimize eddy currents that were continuing to bring large amounts of downstream debris into the stilling basin. The opening of any one gate was not allowed to deviate from that of the other gates by more than 0.3 m (1 ft) initially. This was later reduced to 152 mm (6 in). After this change, the amount of debris removed each year from the basin was drastically reduced, and the rate of abrasion declined; however, nearly 10 years after the repair, the erosion damage had progressed to the same degree that existed prior to the repair.

(4) A materials investigation was initiated prior to the second repair to evaluate the abrasion-erosion resistance of potential repair materials (Holland 1983, 1986). Test results indicated that the erosion resistance of conventional concrete containing a locally available limestone aggregate was not acceptable. However, concrete containing this same aggregate with the addition of silica fume and an HRWRA exhibited high-compressive strengths and very good abrasion-erosion resistance (Figure 8-35). The poor performance of fiber-reinforced concrete cores taken from the original overlay correlates well with the poor performance of the material in the actual structure.

(5) A hydraulic model study of the stilling basin was also conducted prior to the second repair to evaluate potential modifications to the structure that would minimize entry of debris into the basin (Fenwick 1989). Modifications evaluated included a floating boom over the end sill, a downstream dike both longitudinal and lateral, a debris trap, and the paving of the downstream channel. Also, a modification to the upper sluices to provide a manifold to evenly distribute these flows across the stilling basin was investigated. It was suspected that unsymmetrical operation of the lower sluices during the winter months brought material into the basin, while the operation of the upper sluices during the summer months churned the debris around in the basin causing the abrasion-erosion damage. If the debris could not be prevented from entering the basin, then the next approach would be to prevent the swirling discharge from the upper sluices from entering the stilling basin.

(6) Of all the alternatives investigated, it was determined that a debris trap just downstream of the end sill was the least costly with the highest potential for preventing debris from entering the basin. The other alternatives investigated, besides being higher in cost, were subject to possible destruction by spillway flows. The basic purpose behind this plan was to stop the debris from entering the

Figure 8-34. General view of repair operations, Kinzua Dam, 1974

stilling basin by restricting upstream flows and thereby allowing the debris to drop into the 7.6-m (25-ft)-wide trap. The top of the debris trap wall was at the same elevation as the top of the end sill and extended across the stilling basin width. The concrete debris trap wall was an inverted tee wall founded on a varying rock foundation. The debris trap wall was designed to withstand the hydraulic forces during spillway flows.

(7) It was believed that the symmetrical operation of the lower sluice gates would totally prevent the transportation of material into the stilling basin. Therefore, the

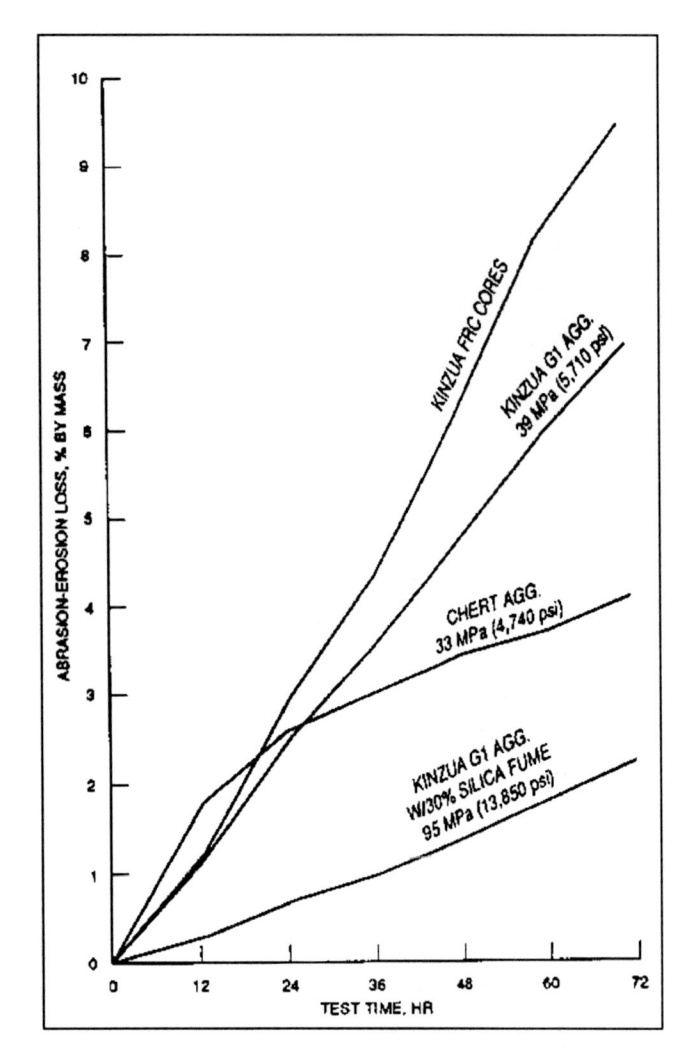

Figure 8-35. Results of selected abrasion-erosion tests, Kinzua Dam

concrete was $353/cu m ($270/cu yd). Construction of a debris trap immediately downstream of the stilling basin end sill was also included in the repair contract. The trap was 7.6 m (25 ft) long with a 3-m (10-ft)-high end sill that spanned the entire width of the basin.

(9) Following dewatering and a thorough cleanup, the transit-mixed silica-fume concrete was pumped into the basin through approximately 46 m (150 ft) of 127-mm (5-in.) pump line (Figure 8-36). The concrete was consolidated with internal vibrators. Once the majority of the concrete was in the form, the screeding operation was started. The two vibrating screeds were connected in tandem with the second following the first at about a 1.5-m (5-ft) interval. Curing compound was applied immediately after the second screed passed over the concrete.

(10) Although the placements generally went well, there was a problem that arose during the construction: cracking of the silica-fume concrete overlay. The widths of cracks were from 0.25 to 0.51 mm (0.01 to 0.02 in.) and usually appeared 2 to 3 days after placement. The cracks were attributed primarily to restraint of volume changes resulting from temperature gradients and, possibly, autogenous shrinkage. Several approaches, including applying insulating blankets over the concrete, saw cutting the slab after placing to establish control joints, and addition of reinforcing steel mat, were attempted to stop or minimize the cracking; however, no solution was found for the cracking problem. The cracks were not repaired because laboratory testing of cracked specimens indicated that fine cracks do not affect abrasion resistance.

(11) Concrete materials and mixture proportions similar to those used in the Kinzua Dam repair were later used in laboratory tests to determine those properties of silica-fume concrete which might affect cracking (McDonald 1991). Tests included compressive and tensile splitting strengths, modulus of elasticity, Poisson's ratio, ultimate strain capacity, uniaxial creep, shrinkage, coefficient of thermal expansion, adiabatic temperature rise, and abrasion erosion. None of these material properties, with the possible exception of autogenous shrinkage, indicated that silica-fume concrete should be significantly more susceptible to cracking as a result of restrained contraction than conventional concrete. In fact, some material properties, particularly ultimate strain capacity, would indicate that silica-fume concrete should have a reduced potential for cracking.

debris trap would serve only as insurance against debris entering the stilling basin for any unanticipated reason. Although prototype studies indicated that velocities were not high enough to bring debris in, the fact remained that debris was removed from the stilling basin annually. Since the operation of the sluice gates could also be controlled by the power company, there was concern that the gates could be accidentally or unintentionally operated in an unsymmetrical manner. Previous studies indicated that unsymmetrical operation for even a few hours would bring in large amounts of debris.

(8) Approximately 1,530 cu m (2,000 cu yd) of silica-fume concrete was used in a 305-mm (12-in.) minimum thickness overlay when the stilling basin was repaired in 1983 (Holland et al. 1986). The bid price for silica-fume

Figure 8-36. Pumping silica-fume concrete into the stilling basin from the top of the right training wall, Kinzua Dam

(12) The potential for cracking of restrained concrete overlays, with or without silica fume, should be recognized. Any variations in concrete materials, mixture proportions, and construction practices that will minimize shrinkage or reduce concrete temperature differentials should be considered. Where structural considerations permit, a bond breaker at the interface between the replacement and existing concrete is recommended.

(13) Four diver inspections were made in the year following the placement of the overlay. During the first inspection (April 1984), very little deterioration was noted and the divers estimated that the curing compound was still in place over 90 percent of the slab. The divers recovered about one and one-half buckets of small gravel and a small reinforcing bar from the basin after the inspection. The second diver inspection was made in August 1984, after a period of discharge through the upper sluices. Erosion to about 13 mm (1/2 in.) in depth was found along some of the cracks and joints. The divers also discovered a small amount of debris and two pieces of steel plating that had been embedded in the concrete around the intake of one of the lower sluices. Because of concern about further damage to the intake, the use of this sluice in discharging flows was discontinued. This nonsymmetrical operation of the structure resulted in the development of eddy currents. Consequently, a third inspection in late August 1984 found approximately 75 cu m (100 cu yd) of debris in the basin. In September 1984, a total of approximately 380 cu m (500 cu yd) of debris was removed from the basin, the debris trap, and the area immediately downstream of the trap. Debris in the basin ranged in size up to more than 305 mm (12 in.) in diameter. Despite these adverse conditions, the silica-fume concrete continued to exhibit excellent abrasion resistance. Erosion along some joints appeared to be wider but remained approximately 13-mm (1/2-in.) deep.

(14) Sluice repairs were completed in late 1984, and symmetrical operation of the structure was resumed. A diver inspection in May 1985 indicated that the condition of the stilling basin was essentially unchanged from the preceding inspection. A very small amount of debris, approximately 0.1 cu m (3 cu ft), was removed from the basin. The debris trap was also reported to be generally clean with only a small amount of debris accumulated in the corners. A diver inspection approximately 3-1/2 years after the repair indicated that the maximum depth of erosion, located along joints and cracks, was about 25 mm (1 in.).

e. Dworshak Dam. A diver inspection in May 1973, about 2 years after the basin became operational, indicated that rubble and materials from construction of the dam had entered the stilling basin and had caused severe damage to the concrete. An inspection in June 1975 indicated that erosion had progressed completely through the 3-m (10-ft)-thick floor slab in some areas. It was estimated that approximately 1,530 cu m (2,000 cu yd) of concrete had been eroded from the stilling basin. This extensive damage was attributed to two factors: (1) large amounts of construction debris deposited in the basin prior to and during initial operation, and (2) unbalanced flow into the basin because of inoperable or faulty gates in the spillway and outlet works.

(1) The stilling basin was dewatered for repair in 1975. An anchored overlay of fiber-reinforced concrete (381-mm (15-in.) minimum thickness) was applied to the basin floor. Flexural and compressive strengths of the fiber concrete mixture were approximately 6 and 55 MPa (860 and 8,000 psi), respectively, at 28 days. Following completion of the overlay, the concrete in the right half of the basin was impregnated with methyl methacrylate monomer (Schrader and Kaden 1976).

(2) In areas where erosion of the original concrete was less than the 381 mm (15-in.) minimum depth specified for the overlay, the design called for removal of the existing concrete to this depth. However, one section of the stilling basin floor and the lower portion of the spillway exhibited only minimal erosion to a maximum depth of about 100 mm (4 in.). Therefore, it was decided to repair both sections, totaling approximately 475 sq m (5,100 sq ft), with an epoxy mortar topping. Several types of epoxy mortar were used to complete this work; the primary one was a stress-relieving material which was slow curing and had a low exotherm. Several problems which were primarily the result of workmanship, weather conditions, and failure to enclose the work area occurred with the epoxy during application. Under the cool conditions that existed, the epoxy mortar probably did not receive full cure before the stilling basin was put back into service.

(3) During the 7 months between completion of repairs and the initial diver inspection, the basin was subjected to a total of 53 days usage (9 days from the spillway gates and 44 days from the regulating outlets). The spillway and outlet gates were operated symmetrically (or very close to it) for all spills. Total flows varied between 59 and 566 cu m/sec (2,100 and 20,000 cu ft/sec), with the majority being on the order of 85 to 283 cu m/sec (3,000 to 10,000 cu ft/sec).

(4) The underwater inspection of the stilling basin by diver identified isolated accumulations of gravel, rebar, and other debris at a number of locations throughout the basin. Since the stilling basin had been completely cleaned following repair and current erosion was not sufficient to expose reinforcement or produce large-size gravel, it was concluded that the material was entering the basin from downstream of the end sill. The inspection indicated no major erosion or damage. The stilling basin walls had a small amount of surface erosion (less than 25 mm (1 in.)). There were several areas at the junction between the floor and wall with erosion up to 76-mm (3-in.) deep. An estimated 25 percent of the surface area of the epoxy mortar had experienced some degree of failure ranging up to 102-mm (4-in.) depths. The fiber-reinforced concrete (both polymerized and non-polymerized) was generally in good condition. In general, the polymer-impregnated side was probably a little better than the nonpolymerized side. There were several areas of erosion in the center of the basin to depths of about 25 mm (1 in.) deep. Joints and open cracks in the entire basin (including the fiber-reinforced concrete) were the most susceptible to damage. Typical joints and open cracks in the fiber concrete had eroded up to about 25-mm (1-in.) deep at the joint and tapered out to the original floor surface within a foot of the joint. Because of the moisture in the joints and cracks during the repair, concrete at joints and cracks was not impregnated.

(5) Four months later (Nov 1976), after some additional usage of the stilling basin, a diver was employed to clean the debris from the basin and provide more information on the condition of the floor. Significant comments resulting from this inspection were that there were large areas of the concrete surface near the center of the basin with grooves 51 to 76 mm (2 to 3 in.) deep. These grooves, in both the polymerized and nonpolymerized fiber-reinforced concrete, were oriented in the direction of flow.

(6) The next diver inspection of the stilling basin in October 1977 indicated that the basin remained clear of debris. Since there had been no spill between inspections, it was concluded that operation of the powerhouse adjacent to the stilling basin would not in itself cause debris to enter the basin. At this point, a hydraulic model study of the stilling basin was initiated.

(7) The 1:50-scale model demonstrated that debris moved upstream into the stilling basin area as a result of flow conditions at the end of the structure when the spillway was in use (U.S. Army Engineer District, Walla Walla 1979). Eddy action at the edges of the outflow

flume moved 16- to 254-mm (5/8- to 10-in.) gravel and cobble near the basin into the runout excavation along the ends of the basin walls only. This material was carried to the upstream edge of the end sill by the roller beneath the flow separating from the sill. Occasionally, the highly turbulent flow lifted from the edge of the sill and the bottom roller swept material into the basin. With time, sizeable amounts of debris accumulated in the basin. The tendency was that the higher the discharge, the greater the movement of material into the basin.

(8) A 6.1-m (20-ft)-high, 1.5-m (5-ft)-thick sill across the basin 3 m (10 ft) upstream from the existing end sill of the same height formed a rock trap that effectively confined debris coming into the basin. Sills of lower heights were investigated but were unsatisfactory. Low walls that extended downstream from the basin walls were also effective in stopping movement of debris into the basin. Sheet-pile walls that were 4 m (13 ft) higher than the downstream channel and extended 15 m (50 ft) downstream, 1.8 m (6 ft) beyond the end of the runout excavation, blocked the movement of debris at the ends of the basin walls. However, a check of prototype site conditions revealed that the area was in rock and therefore unsuited for sheet piles. A second wall-extension plan had low fills 7.6 m (25 ft) long and a minimum of 1.2 m (4 ft) high (Figure 8-37), which might be constructed of tremie concrete. Although lower and shorter, the fills were as effective as the sheet-pile walls in stopping debris movement and confining the bottom roller. The fills were recommended as the best plan for debris control.

(9) Extension of the right training wall was completed in February 1980. The stilling basin was inspected in September 1981 by use of an underwater television camera. No additional damage was observed during the inspection. The stilling basin was generally clean except for a thin skiff of fine material which had settled on the bottom since the spillway was last used in the spring of 1981.

(10) Inspection of the stilling basin was performed by divers with an underwater video camera in October 1983. Several small areas of apparent erosion were identified on the upstream left side of the basin floor. It was difficult to determine whether these areas were, in fact, erosion of the concrete or remnants from previous stilling basin repairs. During past polymerization of concrete surfaces, thin layers of sand adhered to the surface, resulting in a rough texture. This may explain what appeared to be minor erosion damage.

f. Chief Joseph Dam. The stilling basin at Chief Joseph Dam is approximately 280 m (920 ft) wide and 67 m (220 ft) long and is divided into four rows of concrete slabs approximately 20 m (65 ft) wide, 15 m (50 ft) long, and 1.5 m (5 ft) thick. Extensive areas of eroded concrete were discovered in the stilling basin during an underwater inspection in March 1957, 2 years after the project became operative. By 1966, erosion had progressed to maximum depths of approximately 1.8 m (6 ft), with the most severe erosion located in areas between the row of baffles and the end sill. Because of the high cost of dewatering, it was decided to repair the basin slabs underwater using preplaced-aggregate concrete and pumped concrete.

(1) In the preplaced-aggregate concrete operations, concrete buckets containing the coarse aggregate were guided into position and dumped by a diver. Screeds on preset edge forms were then used by the divers for leveling the aggregate to the proper grade before placing the top form panels. Grout pipes were driven the full depth of the coarse aggregate. The grout mixer and pumps were set up on a training wall and grout was first pumped through grout pipes in the deepest area of the placement until a return appeared through the vent pipes surrounding that area. These were plugged with corks when good sanded grout appeared. When grout appeared in the next row of vent pipes, the grout hose was moved to an adjacent pipe and the grout hole plugged. This procedure was continued until the entire form had been pumped. Approximately 61 cu m (80 cu yd) of preplaced-aggregate concrete was required in the repair.

(2) In the pumped concrete operations, concrete was batched at a local supplier and delivered to the site in transit mixers. At the site, the material was placed in a collection hopper on a training wall and delivered from this point through a 254-mm (10-in.) pipe into concrete buckets on a barge for ferrying to the pump sites. From this point the concrete was pumped through a 76-mm (3-in.) hose. The last 0.6 m (2 ft) of this hose was fitted with a metal tube to allow ready insertion of the conduit into the concrete when the surface leveled off. A clamp was located just above the metal tube to provide the required valve action and allow the placement to be controlled by the diver. The concrete mixture produced a workable material that was easily placed and produced a smooth even surface. Surface slopes estimated at 12:1 were achieved. The measured surface area of pumped concrete was 367 sq m (439 sq yd).

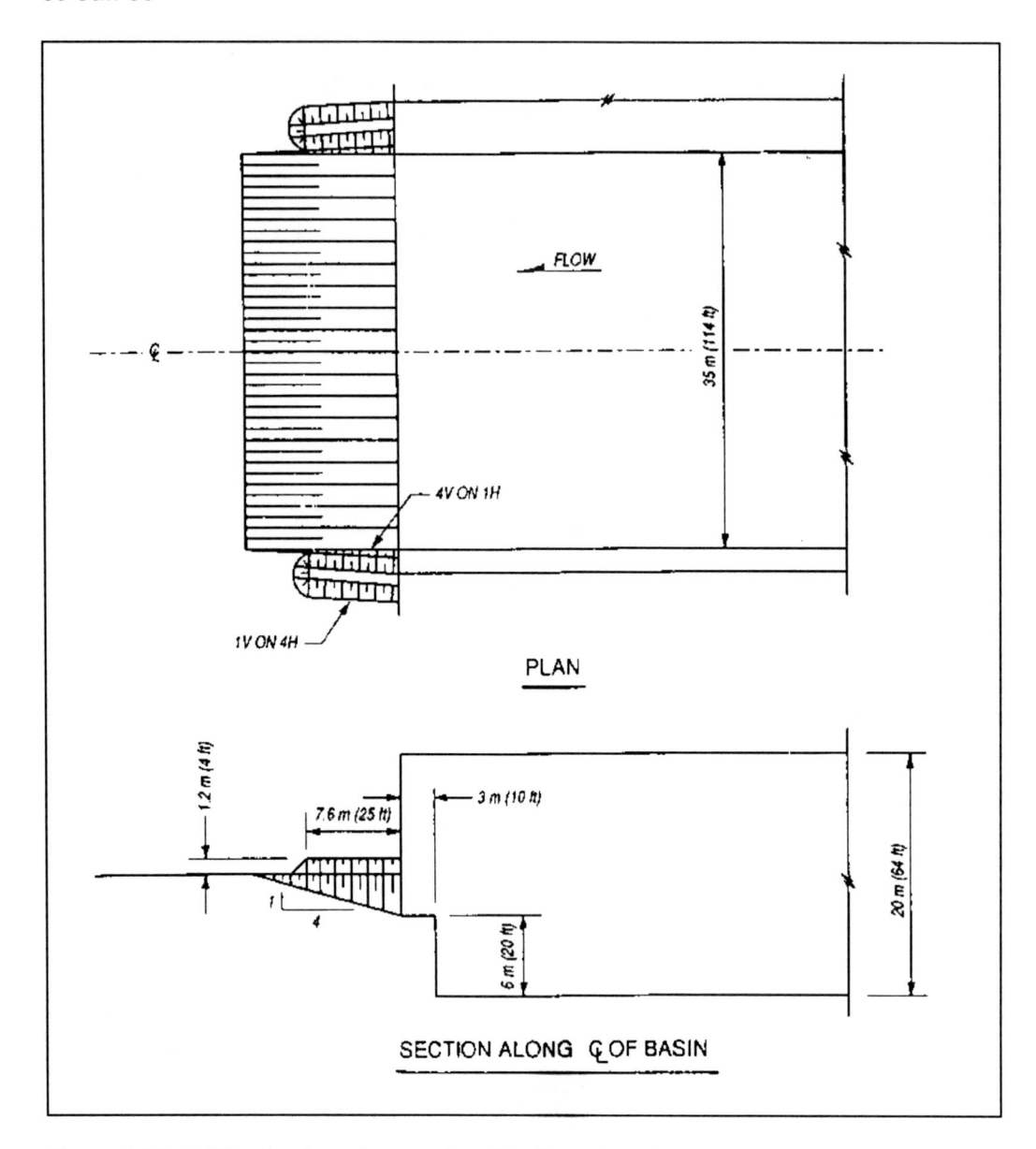

Figure 8-37. Stilling basin wall extension fills, Dworshak Dam

(3) The repairs were accomplished in September through December 1966. During the high-water season in 1967, the repairs were subjected to peak discharges of 12,230 cu m/sec (432,000 cu ft/sec), a flow with a frequency of recurrence of about once in 6 years. A December 1967 inspection of the repairs showed the preplaced-aggregate concrete surfaces to be in excellent condition. The pumped concrete was reported to be in good condition with only minor surface damage noted. The worst damage occurred to the first placement of

pumped concrete where the design mixture was too stiff. A detailed inspection of the stilling basin conducted in 1974 indicated there had been no extensive erosion of the stilling basin since the repairs were made.

g. *Summary.* The repair materials and techniques described in the preceding case histories have been in service for various lengths of time and have been exposed to different operational conditions as well as different levels and durations of flow. This dissimilarity makes any comparison of the relative merits of the various systems difficult at best; however, a number of general trends are apparent.

(1) Materials.

(a) The resistance of steel plate to abrasion erosion is well established; however, it must be sufficiently anchored to the underlying concrete to resist the uplift forces and vibrations created by flowing water. Welding of anchor systems as nearly flush with the plate surface as possible appears more desirable than raised-bolt connections. In any case, the ability of the anchor system, including any embedment material, to perform satisfactorily under the exposure conditions, particularly creep and fatigue, should be evaluated during design of the repair.

(b) It is feasible to use prefabricated panels or cast-in-place concrete in underwater repair of stilling basins. The inherent advantages of each procedure should be considered in those cases where it is extremely difficult and expensive to dewater a structure to make repairs under dry conditions. Prefabricated elements for underwater repair of stilling basins were investigated as part of the REMR research program (Rail and Haynes 1991). Guidance on underwater repair of concrete is given in Section 8-6.

(c) Minimal erosion of the fiber-reinforced grout at Old River Low Sill Structure has been reported. However, the location of this material on a 1-in-7 slope at the end sill should significantly reduce the potential for the presence of abrasive debris on surface of the repair material. The fiber-reinforced concrete failed under the severe abrasion condition at Kinzua Dam. Also, the poor performance of fiber-reinforced concrete cores in abrasion-erosion tests correlated well with the poor performance of the material in the actual structure. After 1-year exposure to limited discharges at Dworshak Dam, fiber-reinforced concrete (both polymerized and nonpolymerized) was eroded to maximum depths of 51 to 76 mm (2 to 3 in.).

(d) An estimated 25 percent of the epoxy mortar at Dworshak Dam had failed, probably because of workmanship, weather conditions, and lack of sufficient curing during construction. No erosion damage to the epoxy mortar was visible at Pomona Dam; however, there were several areas of cracking and loss of bond attributed to improper curing and thermal incompatibility with existing concrete.

(e) Conventional concrete proportioned for 34-MPa (5,000-psi) compressive strength exhibited erosion up to 203 mm (8 in.) deep after less than 1-year exposure at Nolin Lake Dam. In comparison, conventional concrete containing abrasion-resistant trap rock aggregate had general erosion of only 3 to 13 mm (1/8 to 1/2 in.) after 10 years exposure at Pomona Dam. The abrasion resistance of conventional concrete containing the limestone aggregate locally available at Kinzua Dam was unacceptable. However, concrete containing this same aggregate with the addition of silica fume and an HRWRA has exhibited excellent abrasion-erosion resistance. Unlike concretes containing or impregnated with polymers, concretes containing silica fume are economical and are readily transportable and placeable using conventional methods.

(2) Revised Configuration.

(a) The steel modules anchored to the stilling basin slab and the top of the end sill at Old River Low Sill Structure essentially created a sloping end sill. Apparently, any debris discharged through the structure is being flushed from the basin over this fillet, since diver inspections have reported the basin to be free from debris. When the stilling basin at Pomona Dam was dewatered 5 and 10 years after repair, minimal amounts of debris were found in the stilling basin. However, it is difficult to determine if these relatively small amounts of debris were influenced by the addition of a sloping end sill.

(b) A debris trap was provided in the Pomona Dam stilling basin by eliminating the 0.6-m (2-ft)-thick repair overlay in an area between the downstream baffles and the end sill. The debris trap appears to have functioned as planned because most of the debris found in subsequent inspections was located in the trap adjacent to the raised overlay slab. Strong circulatory currents within the stilling basin appear to have negated any effect of the shallow debris trap incorporated into the original repair at Kinzua Dam. Following the adoption of symmetrical discharges, the small amounts of debris present in the stilling basin were located in areas of erosion throughout the basin. The debris trap with a 3-m (10-ft)-high end sill included in the second repair was also unable to prevent downstream debris from entering the basin under the severe hydraulic conditions at Kinzua Dam.

(c) The stilling basin at Nolin Lake Dam was modified during both repairs. In the second repair, modifications were based on the results of a hydraulic model study. These modifications appear to have eliminated previous erosion problems. The extension of the training wall at Dworshak Dam appears to have enhanced the performance of the stilling basin.

(3) Operations.

(a) Model tests of the stilling basin at Pomona Dam verified that severe separation of flow from one sidewall and eddy action within the basin occurred for discharge and tailwater conditions common to the prototype. Also, these tests revealed that the eddy within the basin was capable of generating considerable reverse flow from the exit channel with the potential to transport riprap from the channel into the basin. Based on the model tests, guidance as to the discharge and tailwater relations required to flush debris from the basin was developed. Following implementation of the most practical material and hydraulic modifications, the stilling basin is performing quite well.

(b) Because of the proximity of a pumped-storage power plant on the left abutment and problems from spray, especially during the winter months, the right-side sluices at Kinzua Dam were used most of the time. This usage caused a circulatory current that carried debris from downstream over the end sill, which is below streambed level, into the stilling basin. Under these conditions, an average of 39 cu m (50 cu yd) of debris was removed from the basin during each of three inspections within the 7 months following completion of repairs. At this point a policy of symmetrical sluice operation was initiated, and based on prototype experiments, a table outlining sluice operating procedure for a range of outflow was prepared. Subsequent to the adoption of this revised sluice operation policy, a minimum of debris was removed from the basin and the rate of erosion decreased.

(c) Upon completion of inspection or repair of a dewatered stilling basin, the basin should be flooded in such a manner to prevent material from temporary access roads, cofferdams, etc., from being washed back into the basin.

h. Conclusions.

(a) Conventional concrete with the lowest practical w/c ratio and hard, abrasion-resistant coarse aggregate is recommended for repair of structures subjected to abrasion-erosion damage. Also, silica-fume concrete appears to be an economical solution to abrasion-erosion problems, particularly in those areas where locally available aggregate otherwise might not be acceptable. The abrasion-erosion resistance of repair materials should be evaluated in accordance with ASTM C 1138 prior to application.

(b) In many cases, underwater repair of stilling basins with prefabricated elements, preplaced-aggregate concrete, pumped concrete, or tremie concrete is an economical alternative to dewatering a stilling basin for repairs under dry conditions. Even when a stilling basin is dewatered, it is often difficult to dry existing concrete surfaces because of leaking cracks and joints. Materials suitable for repair of wet concrete surfaces were identified and evaluated as part of the REMR research program (Best and McDonald 1990b).

(c) Additional improvements in materials should continue to reduce the rate of concrete damage caused by erosion. However, until the adverse hydraulic conditions that caused the original damage are minimized or eliminated, it will be difficult for many of the materials currently being used in repair to perform in the desired manner. Prior to major repairs, model studies of the existing stilling basin and exit channel should be conducted to verify the cause(s) of erosion damage and to evaluate the effectiveness of various modifications in eliminating undesirable hydraulic conditions.

(d) In existing structures, releases should be controlled to avoid discharge conditions where flow separation and eddy action are prevalent. Substantial discharges that can provide a good hydraulic jump without creating eddy action should be released periodically in an attempt to flush debris from the stilling basin. Guidance as to discharge and tailwater relations required for flushing must be developed through model/prototype tests. Periodic inspections should be required to determine the presence of debris in the stilling basin and the extent of erosion.

8-4. Concrete Cutoff Walls

Concrete cutoff walls, sometimes referred to as diaphragm walls, are cast-in-place structures used to provide a positive cutoff of the flow of water under or around a hydraulic structure. The decision to construct a concrete cutoff wall is usually not made as a result of deterioration of the concrete in a structure but rather because of flows under or around the structure. Therefore, the decision to construct such a wall should only be made after a thorough program of geotechnical monitoring and review.

a. Application. Concrete cutoff walls have been used in several instances at structures to reduce or eliminate potentially dangerous flows through foundation materials. The walls are typically unreinforced, approximately 0.6 to 0.9 m (2 to 3 ft) thick and are as deep and as long as required by site conditions. At Wolf Creek Dam in Kentucky, the wall was 683 m (2,240 ft) long and contained elements as deep as 85 m (278 ft). In some cases, a cutoff wall has been constructed only after attempts at grouting have been unsuccessful or have not given assurances of completely eliminating flows.

b. Procedure. In general, cutoff walls are constructed as summarized in the following steps. Kahl, Kauschinger, and Perry (1991), Holland and Turner (1980), and Xanthakos (1979) provide additional information on construction of concrete cutoff walls.

(1) A concrete-lined guide trench is constructed along the axis of the wall. This trench is usually only a few feet deep. The concrete provides a working surface on both sides of the wall, helps to maintain the alignment of the wall, and prevents the shoulders of the excavation from caving into the trench.

(2) The excavation is accomplished with appropriate equipment for the site conditions. Usually, the excavation is done as a series of discontinuous segments with the project specifications limiting the amount of excavation that can be open at any time. As segments are excavated and backfilled with concrete, intervening segments are constructed. Excavation equipment includes clam shells, rock drills, and specialty bucket excavators. The excavating is usually done through bentonite slurry to keep the holes open. Slurry preparation, handling, and cleaning are critical aspects of the project. Careful control must be exercised during the excavation of each segment to maintain verticality. If a segment is out of vertical alignment, the next adjacent segment to be placed may not contact the first segment for its entire depth and a gap may exist in the wall.

(3) Concrete is placed in the segments using tremies. (See Section 6-33 for a general description of tremie placement.) Since the wall is expected to provide a complete cutoff, any discontinuities in the concrete may cause serious problems in the performance of the wall. Problems that have been reported have included zones or portions of the wall containing poorly or completely uncemented aggregates. These problems are usually attributable to an improperly proportioned concrete mixture or to poor placement practices. It is extremely important that project personnel be familiar with the required procedures for these placements and that specifications be strictly enforced. The concrete mixture itself is very important, as it is for any tremie placement. The specifications for the concrete should not be based upon a required compressive strength. Instead, the flowability and cohesion of the concrete are critical. Concrete with the proper characteristics may be proportioned using a minimum cement content of 386 to 415 kg/cu m (650 to 700 lb/cu yd) and a maximum w/c of 0.45. Testing as outlined in CRD-C 32 will be of benefit while developing a suitable concrete mixture.

(4) Core drilling should be specified throughout the project as a means of determining the quality of the concrete in place in the wall.

(5) It is an extremely beneficial practice for these projects to require the contractor to place several test panels outside the actual wall area or in noncritical portions of the wall. These test panels will allow for thorough review of the proposed procedures, concrete mixture, and equipment.

8-5. Precast Concrete Applications

The use of precast concrete in repair and replacement of civil works structures has increased significantly in recent years and the trend is expected to continue. Case histories of precast concrete applications in repair or replacement of a wide variety of structures including navigation locks, dams, channels, floodwalls, levees, coastal structures, marine structures, bridges, culverts, tunnels, retaining walls, noise barriers, and highway pavement are described in detail by McDonald and Curtis (1995). Applications of precast concrete in repair of navigation lock walls are described in Section 8-1. Selected case histories of additional precast concrete applications are summarized in the following.

a. Barker Dam. One of the earliest applications of precast concrete panels as stay-in-place forms was at Barker Dam, a cyclopean concrete, gravity structure located near Boulder, CO. The dam is approximately 53 m (175 ft) high with a crest length of 219 m (720 ft). The dam underwent major rehabilitation in 1947 to replace the deteriorated concrete in the upstream face, to correct leakage problems, and to improve the stability of the dam. Deterioration of the concrete on the upstream face of the dam was caused by exposure to approximately 36 years of cycles of freezing and thawing. The reservoir, which is filled in the spring and early summer primarily by melting snow, is drawn down during the winter leaving the upstream face of the dam exposed.

(1) Rehabilitation of the upstream face of the dam consisted of removing the deteriorated concrete, installing precast reinforced-concrete panels over the entire

upstream face, placing coarse aggregate between the dam face and the precast panels, and then grouting the aggregate (Figure 8-38). Several factors influenced the decision to select this repair method, including: (a) repair had to be completed between the time the reservoir was emptied in the fall and filled in the spring, (b) precast panels and aggregate could be placed in severe winter weather conditions, (c) precasting the panels the summer prior to installation and placing them during the winter reduced the potential for later opening of construction joints, (d) a grout with a low cement content could be used for the preplaced aggregate to minimize temperature rise, providing there was a protective shell of high-quality precast concrete, and (e) precast panels were not as expensive as the heavy wooden forms necessary for the placement of conventional concrete.

(2) Resurfacing of the upstream face of the dam required 1,009 precast concrete panels with a total surface area of 7,110 sq m (8,500 sq yd). The reinforced-concrete panels were precast onsite. Each panel was 203 mm (8 in.) thick and most of the panels were 2.1 m (6.75 ft) wide by about 3.7 m (12 ft) long and weighed about 3,630 kg (4 tons). Prior to panel installation, deteriorated concrete was removed from the upstream face of the dam, anchors were installed in the sound concrete, and a stepped footing of conventional concrete was constructed at the base of the dam. The panels were positioned on the dam with a crawler-crane, and dowels embedded in the panel during precasting were welded to anchor bars in the face of the dam. After the joints were grouted, placement of the aggregate was started. Grouting of the preplaced aggregate began when the water elevation was 4.6 m (15 ft) below the crest of the spillway and was completed in about 10 days with almost no interruption.

(3) The panels were erected and coarse aggregate for the preplaced-aggregate concrete was placed concurrently during the period Jan-Apr 1947. Working conditions during this period were generally miserable with bitter

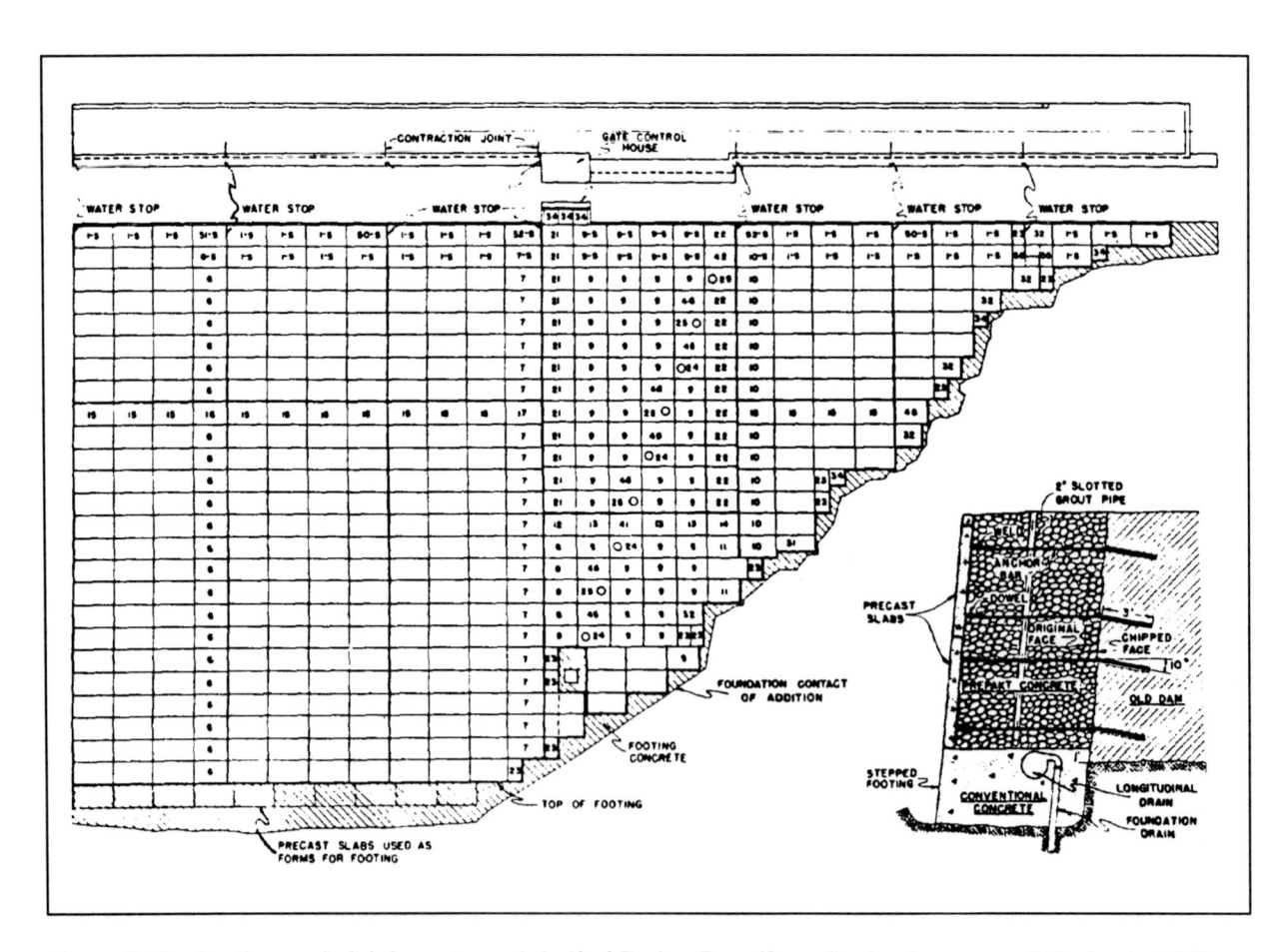

Figure 8-38. Footing and slab layout, south half of Barker Dam (from Davis, Jansen, and Neelands 1948)

cold and high wind velocities. Concrete construction with conventional methods would have been impractical during this period because of the severe weather conditions. The degree of severity of the weather was reflected in rather large daily variations in the rate of panel erection; the average rate was about 12 panels per day with a maximum of 27 panels erected in 1 day.

(4) The quality of the work at Barker Dam is considered to be excellent and the objectives of the rehabilitation program were achieved; however, it is believed that precast panels of much larger size would have resulted in additional economies. With heavy construction equipment, panels up to four times the area could be handled without difficulty and erected at about the same rate as the smaller panels. In addition to reducing the cost of panel erection, larger panels would significantly reduce the total length of joints between panels with a corresponding reduction in the cost of joint treatments.

b. *Gavins Point Dam.* Concrete spalling along the south retaining wall downstream of the powerhouse at Gavins Point Dam was discovered during a diver inspection. Subsequent inspections, which indicated that the spalling was increasing in area and depth, caused concern that the tailrace slab was being undermined. A repair in the dry would have required construction of a cofferdam; however, the cost of building a cofferdam plus the lengthy powerplant outage during construction was too great. Therefore, it was decided that the repairs would be made underwater by divers, working at a depth of approximately 15 m (50 ft).

(1) After the spalled concrete surfaces were cleaned, the voids were filled with preplaced aggregate, covered by precast concrete panels, and grouted (Figure 8-39). The precast concrete stay-in-place forms were anchored to sound concrete to resist uplift caused by the pressure grouting and to improve stability during power plant operation.

(2) The contract specified that the underwater repair was to be completed during a period of 14 consecutive days when the power units were shut down. The contractor used several diving crews so work could continue 24 hours a day. Also, each step of the repair was reviewed on land before being done underwater. As a result, the project was satisfactorily completed 3-1/2 days early.

c. *C-1 Dam.* Lock and Dam C-1 is located on the Champlain Canal near Troy, NY. A two-stage rehabilitation of the dam's seven tainter gate piers was initiated in

1993. In stage I, a cellular cofferdam was constructed to enclose piers 5, 6, and 7. Following dewatering, the counterweights for each tainter gate were removed and placed on temporary supports immediately downstream of the gates. The steel tainter gates were then removed for refurbishing and the existing concrete gate piers were removed down to the original foundation. Precast conrete units were used to reconstruct the gate piers (Figure 8-40).

(1) The concrete mixture used in precasting was proportioned with 13-mm (1/2-in.) maximum size aggregate for a 28-day compressive strength of 48 MPa (7,000 psi). The panels were reinforced with Grade 60, epoxy-coated reinforcing steel. The nose and butt units were cast as individual panels and the remaining pier units consisted of three or four integrated panels. Fabrication tolerances for lengths of the units were 13 mm (1/2 in.), plus or minus, or 3 mm (1/8 in.) per 3 m (10 ft) of length, whichever was greater. Panel thickness and unit width tolerances were 6 mm (1/4 in.).

(2) The precast pier units were transported via tractor-trailer rigs to a launch ramp near the dam. At this point, the loaded rigs were driven onto a barge for transportation to the construction site in the river. The pier units were offloaded with a crane located on the cofferdam. A second crane inside the cofferdam was used to position the pier units. After each tier of units was properly proportioned and aligned, the vertical and horizontal joints were grouted, reinforcement was installed inside the units, and the units were filled with conventional concrete. Eighteen precast units were used in each gate pier. The completed piers are 2.4 m (8 ft) wide, 18.7 m (61.5 ft) long, and 8.7 m (28.5 ft) high. Following completion of the gate piers, the reconditioned tainter gates were reinstalled (Figure 8-41). A similar procedure is currently being used to reconstruct the four remaining gate piers.

d. *Chauncy Run Checkdams.* As part of the overall rehabilitation of the Hornell Local Flood Protection Project, two checkdams were constructed on Chauncy Run. The original plan was to use cast-in-place concrete gravity dams with fully paved stilling basins. However, cost estimates in the 30-percent design submission indicated the use of precast concrete could decrease the cost of the dams by 50 percent. In addition, precasting under controlled plant conditions would assure high quality materials. Precast concrete crib units were used to form the abutments which were then filled with gravel. The checkdams are specially designed precast concrete planks supported by precast concrete posts embedded in the rock

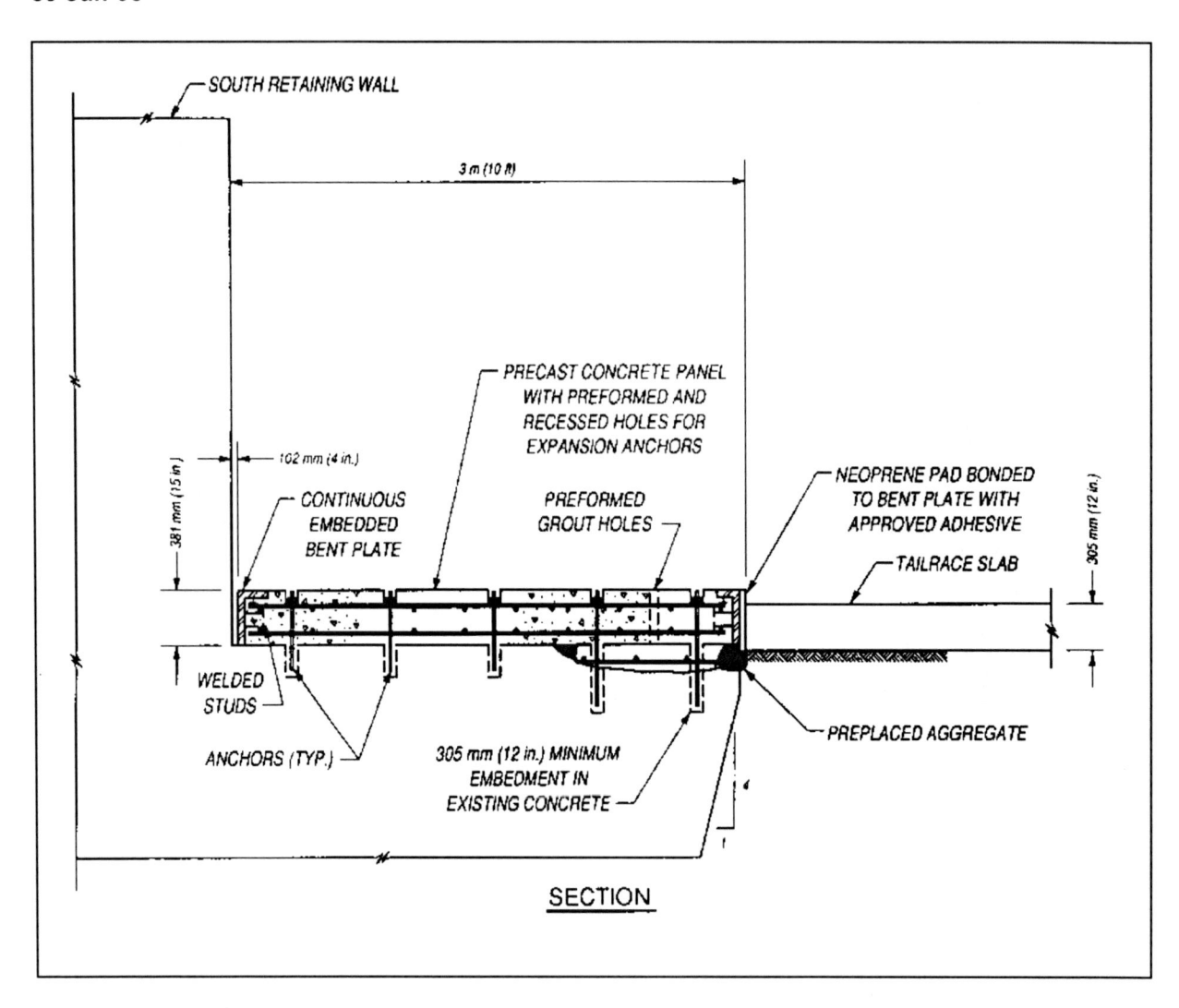

Figure 8-39. Underwater repair of concrete spalling, Gavins Point Dam

foundation (Figure 8-42). The use of precast concrete made it easy to maintain flows during construction, and the appearance of the new structures is well-suited to the site (Figure 8-43).

e. Vischer Ferry Dam. This concrete gravity dam, completed in 1913, is located on the Mohawk River near Albany, NY. The dam consists of two overflow sections with an average height of 12 m (40 ft). The dam was rehabilitated in 1990 as part of an overall project to expand the powerhouse and increase generating capacity. As part of the rehabilitation, the existing river regulating structure was moved to accommodate construction of the expanded powerhouse. The replacement structure is situated perpendicular to the dam so that it discharges from the left side of the new intake. The upstream end of the relocated regulating structure also forms the intake entrance of the forebay. A hydraulic model of the

forebay area showed that head loss and the potential for water separation could be reduced significantly if a contoured pier nose was added at the upstream end of the regulating structure.

(1) The original design for the pier nose was based on cast-in-place concrete inside a dewatered cofferdam. However, the bid cost for the cofferdam alone was $250,000, so the project team reviewed alternatives and decided to use six precast concrete sections stacked vertically with tremie concrete infill. Placement of the precast nose sections (Figure 8-44) and the tremie concrete required approximately 7 working days. The first section was positioned and leveled with jackposts, and sandbags were placed around the perimeter of this segment. Infill concrete was tremied to the top of this section and cured for 4 days. During this time, divers installed guide angles

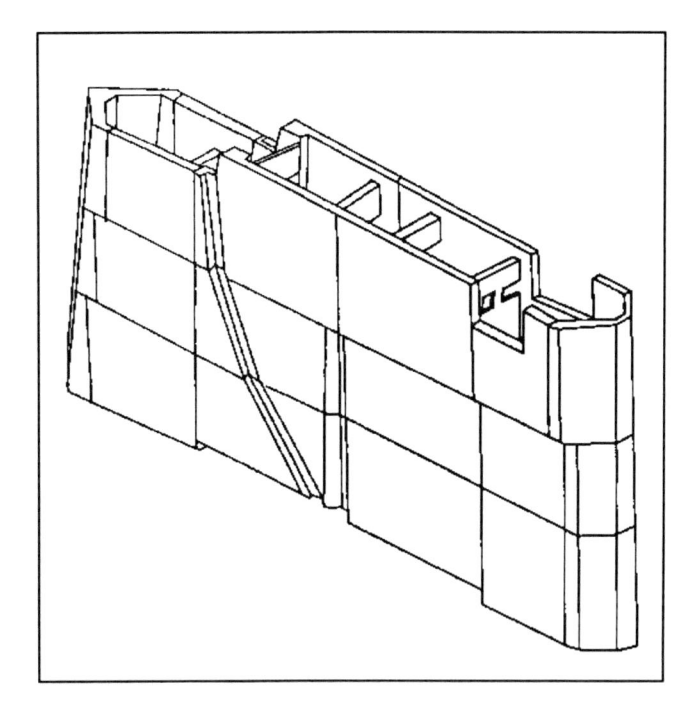

Figure 8-40. Isometric view of gate pier 6, C-1 Dam

Figure 8-41. Completed precast concrete piers with tainter gates reinstalled, C-1 Dam

and reinforcing dowels into the existing pier and spud pipes into the first section. Then, precast sections 2 through 6 were installed, and temporary intermediate connections were installed to resist plastic concrete loads during the second tremie concrete placement. Tremie concrete was placed to within 152 m (6 in.) of the surface at a placement rate that did not exceed 3 m (10 ft) per hr. A cast-in-place concrete cap slab was then formed and placed.

(2) Construction of the rounded pier nose with precast concrete and tremie concrete eliminated the need for a cofferdam and resulted in a savings of $160,000. In addition to reduced construction time and costs, this method effectively eliminated the potentially adverse impact of cofferdam construction on river water quality.

f. Placer Creek Channel. The Placer Creek flood control channel is located in Wallace, ID. For several decades, the channel was repaired and rehabilitated until the channel linings became badly deteriorated and large volumes of debris collected in the channel, reducing its capacity and causing damage to adjacent property.

(1) The contractor elected to use a cast-in-place concrete bottom with precast concrete walls to rehabilitate a 1,128-m (3,700-ft)-long section of the channel (Figure 8-45). Approximately 600 reinforced-concrete panels were precast in a local casting yard. Each panel was 4.6 m (15 ft) long and 3 m (10 ft) high. A 305-mm (12-in.) stub at the bottom of each panel provided continuity of the reinforcement through the corner joint.

(2) By using precast concrete, the contractor was able to reduce (a) rehabilitation time, (b) excavation requirements, (c) costs associated with the forming system, (d) congestion at the restricted project site, and, (e) size of the work force. This use of precast concrete panels resulted in a savings of approximately $185,000.

g. Blue River Channel Project. The Blue River channel modification project was designed to provide flood protection to the Blue River Basin in the vicinity of Kansas City, MO. In an industrial reach of the river, the project consisted of extensive modification of the channel cross section, construction of a floodwall, paving of side slopes, and construction of a 4.6-m (15-ft)-wide by 1.7-m (5.5-ft)-deep low-flow channel for approximately 1.1 km (3,500 ft) of the river. The original design for the low-flow channel consisted of a pair of sheet-pile walls with a cast-in-place concrete strut between the walls. However, there was some concern about this method of construction because of steel smelting slag and other debris embedded in the existing channel. Driving sheet piles through this material could be very expensive or even impossible. Therefore, an alternate design for a precast concrete U-flume was prepared (Figure 8-46).

(1) The estimated cost of construction for the sheet-pile channel was $6,035,000 compared with $8,970,000 to $11,651,000 for the precast concrete channel depending on how the river water was handled during instruction. A

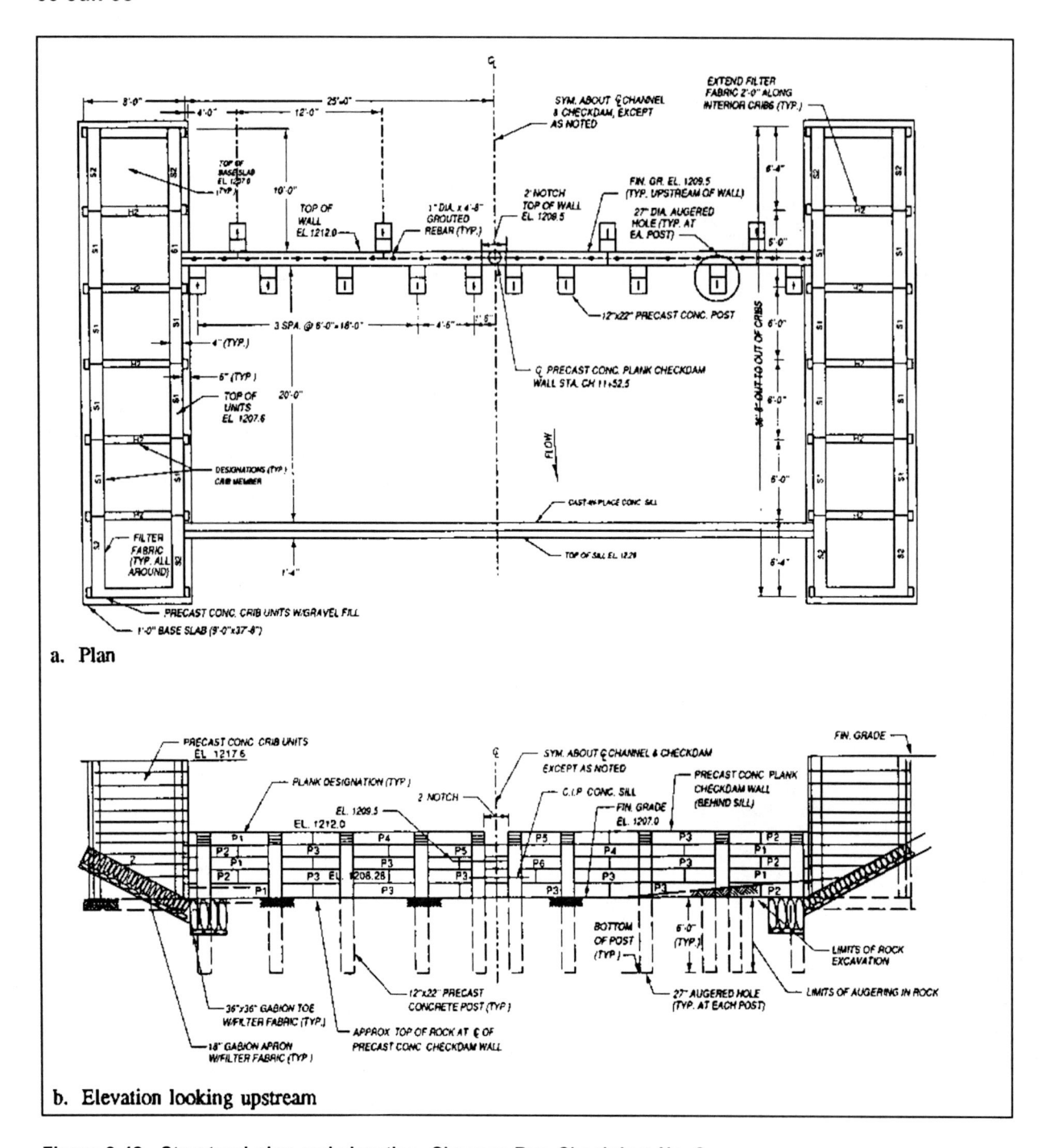

a. Plan

b. Elevation looking upstream

Figure 8-42. Structural plan and elevation, Chauncy Run Checkdam No. 2

Government estimate for the total project cost ($31,852,218) was prepared for the sheet-pile channel only, since it was anticipated to have the lowest total cost.

(2) There were seven bidders on the project and all bids were based on the precast concrete alternate. The bids ranged from $20,835,073 to $37,656,066 with four bids below the Government estimate. The low bidder's estimate for the low-flow channel was $2,000,245, far less than the Government estimate. A comparison of the bids with the Government estimate revealed that selection of the precast concrete alternate was a major factor leading to the bids being considerably lower than the Government

Figure 8-43. Precast concrete checkdams, Chauncy Run

Figure 8-44. Precast pier nose sections ready for underwater installation, Vischer Ferry Dam (Sumner 1993)

estimate. For example, the Government estimate contained $4,753,800 for steel sheetpiling and $898,365 for the concrete in the strut, while the low bid contained just $1,057,500 for the precast concrete sections. Also, the Government estimate for water control was $1,017,000 compared to the low bid of $400,000. The low bidder's estimate for low-flow costs was $2 million, far less than the Government estimate.

(3) During discussions some of the bidders indicated that their estimates showed that, compared to the steel sheet-pile structure, it would be approximately $2 million cheaper to construct the precast concrete U-flume. Their estimates were based on placing 8 to 10 sections per day compared to the Government estimate of 2 sections per day. This difference in production significantly reduced construction time and the cost of water control during construction.

(4) The reinforced-concrete channel sections were precast in 1.5-m (5-ft) lengths which weighed approximately 10,000 kg (11 tons). The concrete in the base slab was placed and cured for 2 days prior to placing the side walls. A concrete compressive strength of 28 MPa (4,000 psi) at 28 days was specified; however, strengths

routinely approached 55 MPa (8,000 psi). Six channel sections were precast daily and stored in the precaster's yard.

(5) The precast sections were shipped, two at a time, by truck to the construction site as required where they were offloaded with a crane and positioned in the channel on a crushed rock subfoundation (Figure 8-47). A total of 700 precast sections were installed with daily placement rates ranging from 12 to 46 sections. The major advantages of precast concrete in this application included low cost, rapid construction, and ease of construction. Also, the use of precast concrete allowed the contractor to divert river water into the channel sections immediately following installation (Figure 8-48). Construction of the low-flow channel was completed in February 1992.

h. Joliet Channel Walls. The walls were constructed in the early 1930's along both banks of the Illinois Waterway through the city of Joliet, IL. The normal pool elevation is 1.2 m (4 ft) below the top of the mass concrete gravity wall on the left bank. The top of the wall is about 2.4 to 9.1 m (8 to 30 ft) higher than the adjacent ground on the landside of the wall.

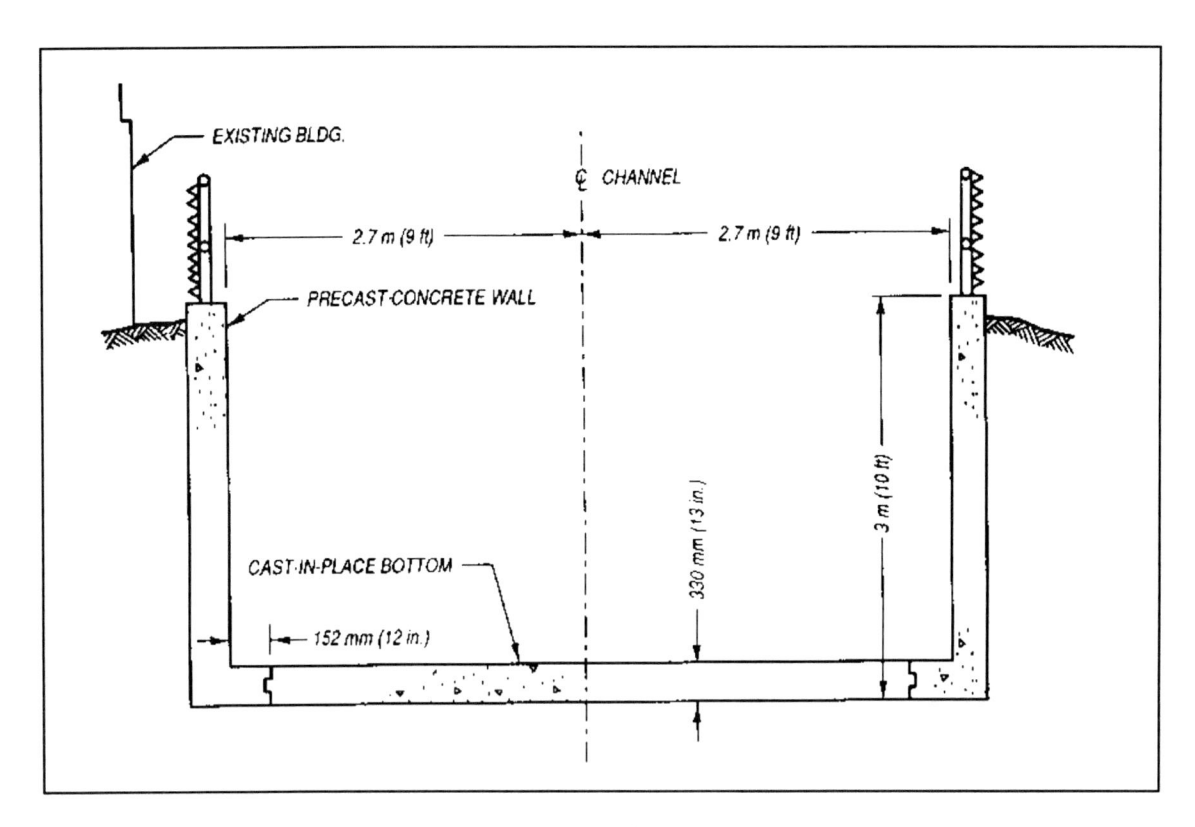

Figure 8-45. Typical channel repair section, Placer Creek (after Hacker 1986)

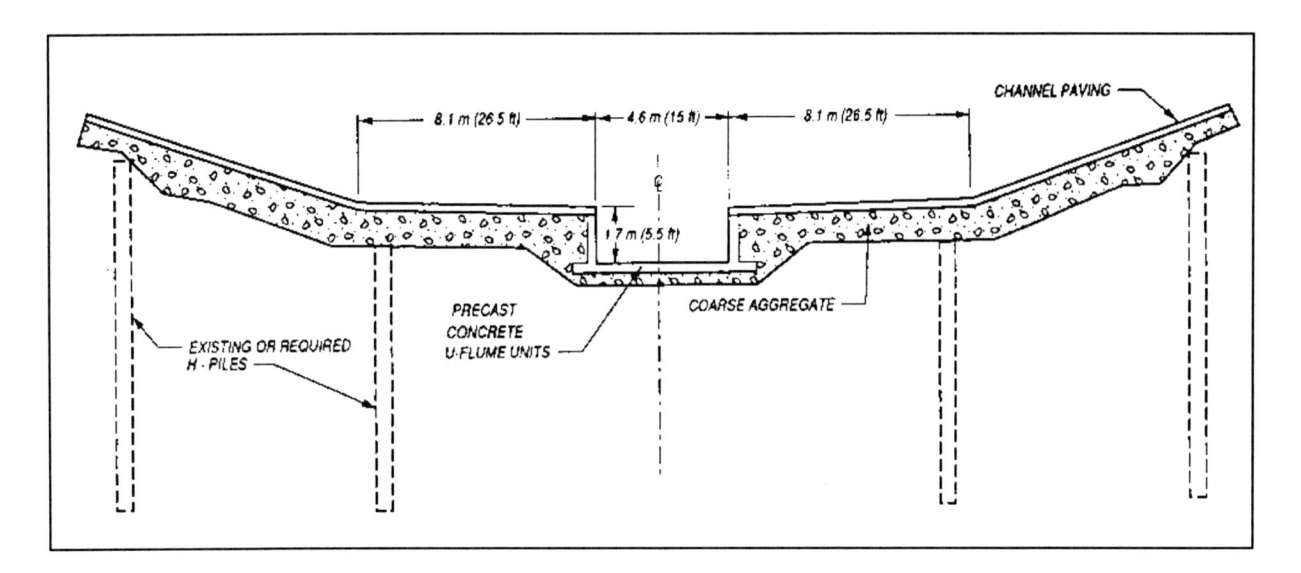

Figure 8-46. Cross section of precast low-flow channel, Blue River Channel (after Mitscher 1992)

(1) A condition survey in 1984 revealed that seepage along monolith and construction joints combined with cycles of freezing and thawing had resulted in extensive deterioration of the exposed concrete to maximum depths of 0.6 m (2 ft). In contrast, those sections of the wall insulated from freezing and thawing by backfill had escaped deterioration. Stability analyses, which considered the depth of deterioration, confirmed that the gravity walls founded on bedrock remained stable. Therefore, it was decided that any repairs should provide aesthetically acceptable insulation for the exposed concrete walls to

Figure 8-47. Channel section delivery and installation, Blue River Channel

Figure 8-48. Diversion of river water through channel sections, Blue River Channel

reduce the potential for additional freeze-thaw deterioration. Earth backfill, the most economical method of providing insulation, was not feasible in all reaches because of buildings near the wall and other right-of-way restrictions. Consequently, a precast concrete panel system that included insulation and drainage provisions was selected for the repair (Figure 8-49).

(2) The reinforced-concrete panels were precast onsite in horizontal lifts. Form oil was used as a bond breaker to allow separation of panels following curing. As many as six panels were cast on top of each other. Typical panels were 254 mm (10 in.) thick, 5.3 m (17.5 ft) high, and 9.1 m (30 ft) long. Each panel had a groove in one end and a partially embedded waterstop in the other.

(3) The installation sequence began with soil excavation to accommodate the panel footing and a perforated pipe drainage system. A 76-mm (3-in.)-diam pipe was installed vertically in the cast-in-place footing on 4.6-m (15-ft) centers. The pipes were used to collect any water seeping through joints in the wall and convey it into the subsurface drainage system.

(4) A 10-mil-thick polyethylene vapor barrier was placed on the existing wall surface; deteriorated concrete was not removed prior to the repair. A 51-mm (2-in.) thickness of extruded insulation was placed on the vapor barrier. The insulation was then covered with another polyethylene vapor barrier. A crane was then used to lift the panels from the horizontal beds to the wall. The panels were positioned so that the groove accommodated the waterstop from the adjacent panel. Expansion-joint material was placed between the panels, and nonshrink grout was placed in the groove to encapsulate the waterstop. A cast-in-place concrete cap on top of the precast panels minimizes moisture intrusion into the repair.

(5) The total length of the precast concrete repair was 349 m (1,145 ft). The contractor's bid price for the wall repair including drainage system, concrete footing, anchors, insulation, vapor barriers, and concrete cap was $207 per sq m ($19 per sq ft). The economical, aesthetically pleasing repair (Figure 8-50) was completed in August 1987 and the repair continues to perform satisfactorily.

i. Ulsterville Bridge. Because this bridge spans a Class 1 trout stream in Ulsterville, NY, it was necessary to minimize onsite construction activities. Consequently,

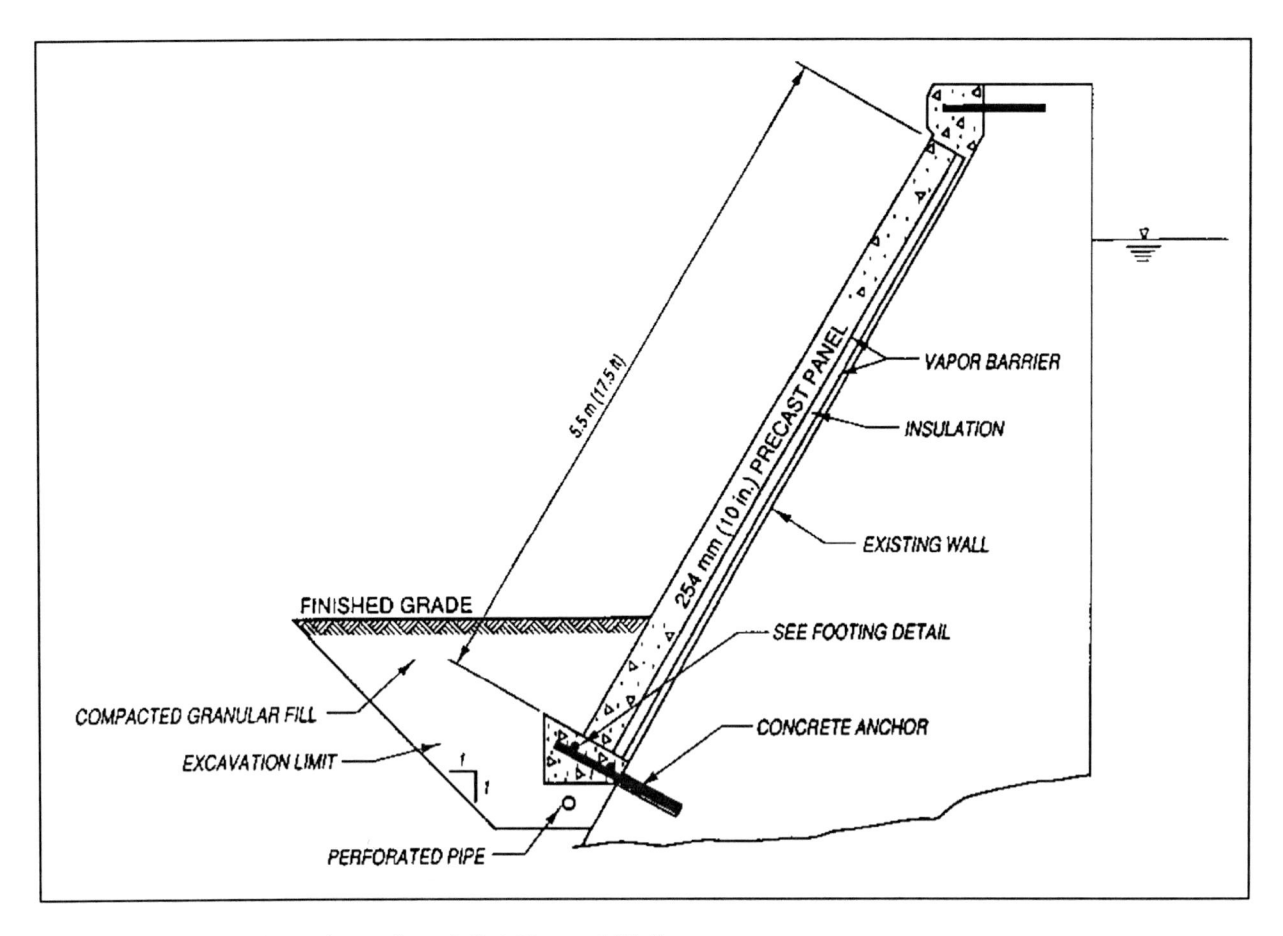

Figure 8-49. Typical repair section, Joliet Channel Wall

precast concrete components were used in September 1989 to construct the bridge abutments and deck.

(1) Precast modules of reinforced concrete stacked on a base slab were used to construct the abutments (Figure 8-51). The modular units were precast in "startup" wood molds, which resulted in some minor fitting problems in the field; however, adjustments were made without the construction being interrupted. Once the modules were properly positioned and aligned, the back cavities in the units were filled with select granular backfill. Vertical reinforcing steel was inserted into the front cavities and these cavities were filled with cast-in-place concrete to form a sealed, monolithic-like front wall (Figure 8-52).

(2) Once the abutments were completed, the precast deck was installed. The composite deck units were precast upside down in forms suspended from wide-flange steel girders. Stud shear connectors were welded to the girders. This technique uses the weight of the forms and the concrete to produce a prestressed effect on the girders.

Another result of the upside down casting is that the densest, least permeable concrete is on the wearing surface. When the cured deck units are turned over, the concrete is precompressed which increases its resistance to cracking. The deck units were placed with a crane; steel diaphragms were installed between the units, and then all longitudinal joints were sealed with grout. The bridge was ready for traffic as soon as construction was complete (Figure 8-53).

(3) Since all components were precast, a small local contractor was able to complete the project within a few days. The cost of the bridge was very competitive with alternate construction procedures and there was minimal environmental impact on the existing trout stream.

j. Summary. The use of precast concrete in repair and replacement of civil works structures has increased significantly in recent years and this trend is expected to continue. A review of these applications shows that, compared with cast-in-place concrete, precasting offers a

Figure 8-51. Top view of precast abutment prior to backfilling, Ulsterville Bridge

Figure 8-50. Joliet Channel Wall before and after repair

number of advantages including ease of construction, rapid construction, high quality, durability, and economy.

(1) Precasting minimizes the impact of adverse weather. Concrete fabrication in a precaster's plant can continue in winter weather that would make onsite cast-in-place concrete production cost prohibitive or impossible. Also, precast concrete can be installed underwater and in weather conditions where construction with conventional cast-in-place concrete would be impractical.

(2) Concentrating construction operations in the precaster's plant significantly reduces the time and labor required for onsite construction. Reducing onsite construction time is a major advantage in repair of hydraulic structures such as locks and dams where delays and shutdowns can cause significant losses to the users and

Figure 8-52. Front view of precast abutment prior to placing concrete cap, Ulsterville Bridge

owners. Also, rapid construction minimizes the potential for adverse environmental impact in the vicinity of the project site.

(3) Although the quality and durability of precast concrete is not necessarily better than concrete cast in place at the project site, a qualified precaster usually has the advantages of a concentrated operation in a fixed plant with environmental control; permanent facilities for forming, batching, mixing, placing, and curing; well

Figure 8-53.   Precast abutment and deck units, Ulsterville Bridge

established operating procedures, including strict quality control; and, personnel with experience in the routine tasks performed on a daily basis. Also, precasting makes it possible to inspect the finished product prior to its incorporation into the structure.

(4) Ease of construction, rapid construction, and repetitive use of formwork all contribute to lower construction costs with precast concrete. Also, underwater installation of precast concrete eliminates the significant costs associated with dewatering of a hydraulic structure so that conventional repairs can be made under dry conditions. As the number of qualified precast suppliers continues to increase and as contractors become more familiar with the advantages of precast concrete, it is anticipated that the costs of precast concrete will be further reduced.

8-6. Underwater Repairs

Dewatering a hydraulic structure so that repairs can be made under dry conditions is often (a) difficult, and in some cases, practically impossible, (b) disruptive to project operations, and (c) expensive. For example, costs to dewater a stilling basin can exceed $1 million, and the average cost to dewater is more than 40 percent of the total repair cost (McDonald 1980). Consequently, studies were conducted as part of the REMR research program to develop improved materials and techniques for underwater repair of concrete. These studies included (a) identification of methods and equipment for underwater cleaning and inspection of concrete surfaces, (b) evaluation of materials and procedures for anchor embedment in

hardened concrete under submerged conditions, (c) development of improved materials and techniques for underwater placement of freshly mixed concrete, and (d) development of prefabricated elements for underwater repair. Underwater inspection of concrete surfaces is covered in Section 2-4. Results of the remaining studies are summarized in the following.

a. Surface preparation. All marine growth, sediments, debris, and deteriorated concrete must be removed prior to placement of the repair material. This surface preparation is essential for any significant bond to occur between the repair material and the existing concrete substrate. Equipment and methods specifically designed for underwater excavation and debris removal are available. Also, a wide variety of underwater cleaning tools and methodologies have been designed specifically for cleaning the submerged portions of underwater structures. The advantages and limitations of each are described in detail by Keeney (1987) and summarized in the following.

(1) Excavation. The three primary methods for excavation of accumulated materials, such as mud, sand, clay, and cobbles, include: air lifting, dredging, and jetting. Selection of the best method for excavation depends on several factors: the nature of the material to excavated; the vertical and horizontal distances the material must be moved; the quantity of the material to be excavated; and, the environment (water depth, current, and wave action). General guidance on the suitability of the various excavation methods is given in Figure 8-54.

(a) Air lifts should be used to remove most types of sediment material in water depths of 8 to 23 m (25 to 75 ft).

(b) Jetting and dredging techniques, or combinations thereof, are not limited by water depth.

(2) Debris removal. The primary types of debris that accumulate in hydraulic structures, such as stilling basins, include cobbles, sediment, and reinforcing steel. Cobbles and sediment can be removed with one of the excavation techniques discussed in the previous section. Removal of exposed reinforcing steel often requires underwater cutting of the steel. There are three general categories of underwater steel cutting techniques. The two most common techniques are mechanical and thermal.

(a) Equipment used for mechanical cutting includes portable, hydraulically powered shears and bandsaws.

Excavation Factor	Excavation Method		
	Air Lift	Jet	Dredge
Type of seabed material	mud, sand, silt, clay, cobbles	mud, sand, silt, clay	mud, sand, silt, clay
Water depth	8 to 23 m (25 to 75 ft)	unlimited	unlimited
Horizontal distance material moved	short	short	short to long
Vertical distance material moved	short to long	short	short to medium
Quantity of material excavated	small to large	small to medium	small to medium
Local current	not required	required	not required
Topside equipment required	compressor	pump	pump
Shipped space/weight	large	small	medium

Figure 8-54. Guidance on excavation techniques (Keeney 1987)

(b) Three thermal techniques are recommended for underwater cutting: oxygen-arc cutting, shielded-metal-arc cutting and gas cutting. Oxygen cutting is the preferred technique for Navy Underwater Construction Team diver operations.

(c) The technology exists for underwater abrasive-jet cutting systems, and commercially available equipment is evolving.

(3) Cleaning. There are three general types of cleaning tools: hand tools, powered hand tools, and, self-propelled cleaning vehicles. Hand tools include conventional devices such as scrapers, chisels, and wire brushes. Powered hand tools include rotary brushes, abrasive discs, and water-jet systems. Self-propelled cleaning vehicles are large brush systems that travel along the work surface on wheels. The types of cleaning tools recommended for different types of material, fouling, and surface area are shown in Figure 8-55.

(a) On large and accessible concrete surfaces, a self-propelled vehicle can be used to quickly and effectively remove light to moderate marine and freshwater fouling.

For areas that are not large enough to justify the use of a self-propelled vehicle, hydraulically powered hand tools, such as rotary cutters, can efficiently remove all fouling from concrete surfaces.

(b) A high-pressure waterjet is the best tool to use in obstructed or limited access areas. A high-pressure, high-flow system can be used to remove most types of moderate to heavy fouling. A high-pressure, low-flow system may be required to clean an area that is difficult or impossible to reach with a high-flow system because of the retrojet.

(c) Because of their low cleaning efficiency, hand tools should be used only where there is light fouling or spot cleaning is to be done in limited areas.

b. Anchors. Repairs are often anchored to the existing concrete substrate with dowels. Anchors are particularly necessary in areas where it is difficult to keep the concrete surface clean until the repair is placed. Anchor systems are categorized as either cast-in-place (anchors installed before the concrete is cast) or

Fouling	Size	Material		
		Concrete	Steel	Timber
Light	Massive	Self-propelled vehicles		Waterjets
	Large	Waterjets and hand-held power tools*		
	Limited Access	High-pressure waterjets		
Moderate	Massive	Self-propelled vehicles		Waterjets/ power tools
	Large	Power tools/ waterjets	**	Power tools/ waterjets
	Limited Access	High-pressure waterjets		
Heavy	Massive	Self-propelled vehicles		N/A
	Large	Power tools	Power tools/ waterjets	Power tools
	Limited Access	High-pressure waterjets		

Notes:

* Hand tools for limited spot cleaning of light and loose fouling.
** Abrasive waterjets for paint removal or bare metal finish on steel structures.

Figure 8-55. Guidance on cleaning tools (Keeney 1987)

postinstalled (anchors installed in holes drilled after the concrete has hardened). Since most of the anchors used in concrete repair are postinstalled, anchors can be classified as either grouted or expansion systems.

(1) Expansion anchors are designed to be inserted into predrilled holes and then expanded by either torquing the nut, hammering the anchor, or expanding into an undercut in the concrete. These anchors transfer the tension load from the anchor to the concrete through friction or keying against the side of the drill hole. Detailed descriptions of the various types of expansion anchors are included in ACI 355.1R.

(2) Grouted anchors include headed or headless bolts, threaded rods, and deformed reinforcing bars. They are embedded in predrilled holes with either cementitious or polymer materials. Cementitious materials include portland-cement grouts, with or without sand, and other commercially available premixed grouts. Polymer materials are generally two-component compounds of polyesters, vinylesters, or epoxies. These resins are available in four forms: tubes or "sausages," glass capsules, plastic cartridges, or bulk. Setting times for polymer materials are temperature dependent and can vary from less than a minute to several hours depending on the formulation.

(3) The effectiveness of neat portland-cement grout, epoxy resin, and prepackaged polyester resin in embedding anchors in hardened concrete was evaluated under a variety of wet and dry installation and curing conditions (Best and McDonald 1990b). Pullout tests were conducted at eight different ages ranging from 1 day to

32 months. Creep and durability tests were also conducted.

(a) Beyond 1 day, all pullout strengths were approximately equal to the ultimate strength of the reinforcing-bar anchor when the anchors were installed under dry conditions, regardless of the type of embedment material or curing conditions. With the exception of the anchors embedded in polyester resin under submerged conditions, pullout strengths were essentially equal to the ultimate strength of the anchor when the anchors were installed under wet or submerged conditions. The overall average pullout strength of anchors embedded in polyester resin under submerged conditions was 35 percent less than the strength of similar anchors installed and cured under dry conditions. The largest reductions in pullout strength, approximately 50 percent, occurred at ages of 6 and 16 months. Although the epoxy resin performed well in these tests when placed in wet holes, it should be noted that the manufacturer does not recommend placement under submerged conditions.

(b) Creep tests were conducted by subjecting pullout specimens to a sustained load of 60 percent of the anchor-yield strength and periodically measuring anchor slippage at the end of the specimen opposite the loaded end. After 6 months under load, anchors embedded in portland-cement grout and epoxy resin that were installed and tested under wet conditions exhibited low anchor slippage, averaging 0.071 and 0.084 mm (0.0028 and 0.0033 in.), respectively, or two to four times higher than results under dry conditions. Anchors embedded in polyester resin, installed and cured under submerged conditions, exhibited significant slippage; in fact, in one case the anchor pulled completely out of the concrete after 14 days under load. After 6 months under load, the two remaining specimens exhibited an average anchor slippage of 2.09 mm (0.0822 in.), approximately 30 times higher than anchors embedded in portland-cement grout under the same conditions.

(c) Long-term durability of the embedment materials was evaluated by periodic compressive strength tests on 51-mm (2-in.) cubes stored both submerged and in laboratory air. After 32 months, the average compressive strength of polyester-resin and epoxy-resin specimens stored in water was 37 and 26 percent less, respectively, than that of companion specimens stored in air. The strength of portland-cement grout cubes stored in water averaged 5 percent higher than that of companion specimens stored in air during the same period.

(4) The performance of anchors embedded in vinyl-ester resin, prepackaged in glass capsules, was also evaluated under dry and submerged conditions (McDonald 1989). Pullout tests were conducted at four different ages ranging from 1 to 28 days. The tensile capacity of anchors embedded under submerged conditions was approximately one-third that of similar anchors embedded in dry holes.

(5) The reduced tensile capacity of anchors embedded in concrete under submerged conditions with prepackaged polyester-resin and vinylester-resin cartridges is primarily attributed to the anchor installation procedure. Resin extruded from dry holes during anchor installation was very cohesive, and a significant effort was required to obtain the full embedment depth. In comparison, anchor installation required significantly less effort under submerged conditions. Also, the extruded resin was much more fluid under wet conditions, and the creamy color contrasted with the black resin extruded under dry conditions. Although insertion of the adhesive capsule or cartridge into the drill hole displaces the majority of the water in the hole, water will remain between the walls of the adhesive container and the drill hole. Insertion of the anchor traps this water in the drill hole and causes it to become mixed with the adhesive, resulting in an anchor with reduced tensile capacity. An anchor-installation procedure that eliminates the problem of resin and water mixing in the drill hole is described by McDonald (1990).

(a) In the revised installation procedure (Figure 8-56), a small volume of adhesive was injected into the bottom of the drill hole in bulk form prior to insertion of the adhesive capsule. This injection was easily accomplished with recently developed paired plastic cartridges (Figure 8-57) which contained the vinylester resin and a hardener. The cartridges were inserted into a tool similar to a caulking gun which automatically dispensed the proper material proportions through a static mixing tube directly into the drill hole. Once the injection was completed, insertion of a prepackaged vinylester-resin capsule displaced the remainder of the water in the drill hole prior to anchor insertion and spinning.

(b) Anchors installed with the revised procedure exhibited essentially the same tensile capacity under dry and submerged conditions. At 3-mm (0.1-in.) displacement, the tensile capacity of vertical anchors installed with the revised procedure under submerged conditions averaged more than three times greater than that of similar anchors installed with the original procedure. The

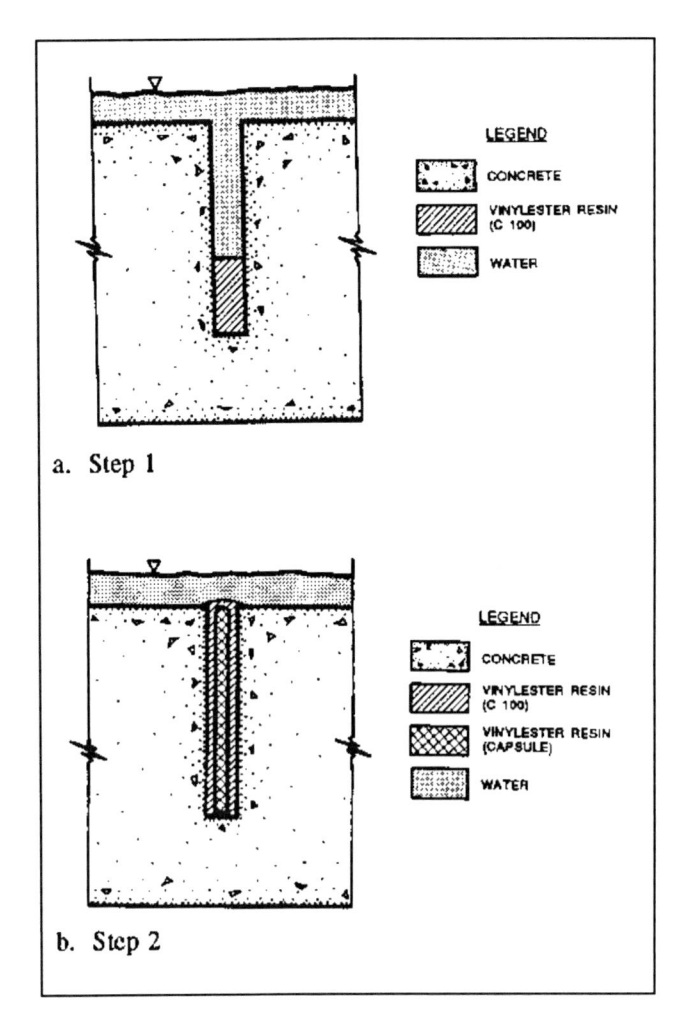

LEGEND

CONCRETE

VINYLESTER RESIN
(C 100)

WATER

a. Step 1

LEGEND

CONCRETE

VINYLESTER RESIN
(C 100)

VINYLESTER RESIN
(CAPSULE)

WATER

b. Step 2

Figure 8-56. Two-step procedure for anchor installation under submerged conditions

ultimate tensile capacity of anchors installed under submerged conditions was near the yield load of the anchors. Also, the difference in tensile capacity between horizontal anchors installed under dry and submerged conditions was less than 2 percent.

(6) Epoxy resins were not prepackaged in "sausage" type cartridges because insertion and spinning of the anchor did not provide adequate mixing. However, with development of the coaxial or paired disposable cartridges with static mixing tubes, a number of suppliers are presently marketing epoxies for anchor embedment under submerged conditions. Also, some suppliers contend that insertion of the prepackaged capsule in the second step of the two-step installation procedure can be eliminated by injecting additional epoxy. However, preliminary results of current tests indicate that anchors installed by epoxy injection alone perform very poorly.

(7) A "nonshrink cementitious anchor cartridge" was recently introduced on the market. According to the supplier, the cartridge contains a fast-setting cementitious compound encased in a unique envelope, which when immersed in water will allow controlled wetting of the contents, forming a thixotropic grout. Results of preliminary tests on anchors installed with this system under submerged conditions are very favorable.

(8) Pending completion of current tests, the two-step anchor installation procedure (Figure 8-56) should be followed when prepackaged vinylester resin is to be used as an embedment material for short (less than 381-mm (15-in.) embedment length) steel anchors in hardened

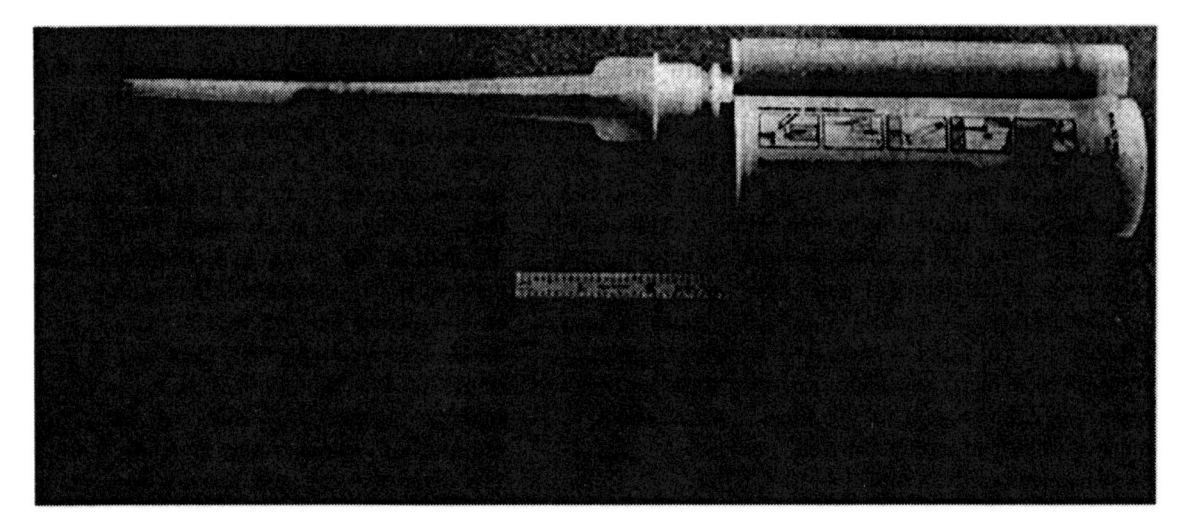

Figure 8-57. Paired disposable cartridges and static mixing tube

concrete under submerged conditions. Similar anchors embedded in neat portland-cement grout exhibit excellent performance when the grout is allowed to cure for a minimum of 3 days prior to loading. The ability of the anchor system including any embedment material, to perform satisfactorily under the exposure conditions, particularly creep and fatigue, should be evaluated during design of the repair.

c. Materials. Cast-in-place concrete and prefabricated elements of concrete and steel have been used successfully in underwater repair of hydraulic structures. Each material has inherent advantages and limitations which should be considered in design of a repair for specific project conditions.

(1) Cast-in-place concrete. Successful underwater concrete placement requires that the fresh concrete be protected from the water until it is in place and begins to stiffen so that the cement and other fines cannot wash away from the aggregates. This protection can be achieved through proper use of placing equipment, such as tremies and pumps (ACI 304R, and Gerwick 1988). Also, the quality of the cast-in-place concrete can be enhanced by the addition of an antiwashout admixture (AWA) which increases the cohesiveness of the concrete (Khayat 1991, Neeley 1988, and Neeley, Saucier, and Thornton 1990). The purpose, types, and functions of AWA's for concrete used in underwater repairs is described in REMR Technical Note CS-MR-7.2 (USAEWES 1985f).

(a) Concrete mixtures for underwater placement must be highly workable and cohesive. The degree of workability and cohesiveness can vary somewhat depending upon the type of placing equipment being used and the physical dimensions of the placement area. For example, massive and confined placements, such as cofferdams, or bridge piers, can be completed with a conventional tremie concrete mixture that is less workable (flowable) and cohesive than a mixture for a typical repair where the concrete is placed in relatively thin sections with large surface areas. An AWA should be used to enhance the cohesiveness of concrete that must flow laterally in thin lifts for a substantial distance. Silica fume should be used to enhance the hardened properties of concrete subjected to abrasion-erosion. The addition of silica fume will also increase the cohesiveness of the fresh concrete mixture. Guidance on proportioning concrete mixtures for underwater placement is given in EM 1110-2-2000.

(b) The tremie method has been successfully used for many years to place concrete underwater (ACI 304R).

The tremie pipe must be long enough to reach from above water to the location underwater where the concrete is to be deposited. Concrete flows through the tremie, the lower end of which is embedded in a mound of the fresh concrete so that all subsequent concrete flows into the mound and is not exposed directly to the surrounding water. Tremie concrete mixtures must be fully protected from exposure to water until in place. AWA's can be used in tremie concrete but are not necessary, although their use will enhance the cohesiveness and flowability of the concrete. If an AWA is used, embedment of the tremie in the fresh concrete is still desirable, but some exposure of the fresh concrete to water may be permitted. Free-fall exposure of the concrete through water is not recommended under any conditions. Usually the tremie is maintained in a vertical position. However, recent research has indicated that inclining the tremie to approximately 45 deg is effective when placement conditions require the concrete to flow laterally several feet in thin layers (Khayat 1991).

(c) The Hydrovalve method and the Kajima Double Tube tremie method (Gerwick 1988) are each variations of the traditional tremie. Each uses a flexible hose that collapses under hydrostatic pressure and thus carries a controlled amount of concrete down the hose in slugs. This slow and contained movement of the concrete helps to prevent segregation and is particularly useful in placing conventional tremie concrete. These methods are reliable, inexpensive, and can be used by any contractor with personnel experienced in working underwater.

(d) In recent years, pumping has become preferable to the tremie method for placing concrete underwater. There are fewer transfer points for the concrete, the problems associated with gravity feed are eliminated, and the use of a boom permits better control during placement. These advantages are especially important when concrete is being placed in thin layers, as is the case in many repair situations. A concrete pump was effectively used for underwater repair of erosion damage at Red Rock Dam (Neeley and Wickersham 1989). A diver controlled the end of the pump line, keeping it embedded in the mass of newly discharged concrete and moving it around to completely fill the repair area. The effects of the AWA used were apparent; even though the concrete had a slump of 229 mm (9 in.), it was very cohesive. The concrete pumped very well and, according to the diver, self-levelled within a few minutes following placement. The diver also reported that the concrete remained cohesive and exhibited very little loss of fines on the few occasions when the end of the pump line kicked out of the concrete. The total cost of the repair was $128,000.

In comparison, estimated costs to dewater alone for a conventional repair ranged from $1/2 to $3/4 million.

(e) Pneumatic valves attached to the end of a concrete pump line permit better control of concrete flow through the lines and even allow termination of the flow to protect the concrete within the lines and to prevent excessive fouling of the water while the boom is being moved. Some units incorporate a level detector to monitor the concrete placement. This method is considered to be one of the most effective for placing concrete underwater (Gerwick 1988). Guidance on pumped concrete is given in EM 1110-2-2000. Also, ACI 304.2R is an excellent reference.

(f) Preplaced-aggregate concrete is an effective method for repairing large void areas underwater. The coarse aggregate is enclosed in forms, and grout is then injected from the bottom of the preplaced aggregate. To prevent loss of fines and cement at the top of the repair, venting forms have a permeable fabric next to the concrete, backed with a wire mesh, and supported by a stronger backing of perforated steel and plywood. The pressure generated by the grout beneath the forms necessitates doweling to hold down the forms. The grout must have a high fluidity to ensure complete filling of voids. Use of an AWA will lessen the need for the protective top form. Guidance on preplaced-aggregate concrete is given in EM 1110-2-2000 and ACI 304.1R.

(g) Some techniques have been developed, such as buckets, skips, and tilting pallet barges, to place stiff, highly cohesive concrete by dropping the concrete through several feet of water, thus requiring the use of a large amount of AWA. However, until these methods have been fully developed and proved effective, there is a substantial risk that the end result will be a poor quality concrete placement. Therefore, placing concrete by allowing it to free-fall through several feet of water is not recommended.

(2) Prefabricated elements. Precast concrete panels and modular sections have been successfully used in underwater repair of lock walls (Section 8-1), stilling basins, and dams (Section 8-5). Prefabricated steel panels have been used underwater as temporary and permanent top forms for preplaced-aggregate concrete. Also, prefabricated steel modules have been used in underwater repair of erosion damage. Rail and Haynes (1991) concluded that it is feasible to use prefabricated steel, precast concrete, or composite steel-concrete panels in underwater repair of stilling basins. Each material has inherent advantages, and the following factors should be considered in designing panels for a specific project.

(a) Abrasion resistance. It is the opinion of Rail and Haynes (1991) that the abrasion resistance of steel is far superior to that of concrete; however, no data were given to substantiate this claim. Recent tests by Simons (1992) indicate that the widths and depths of abrasion of concrete mixtures with an average compressive strength of 75 MPa (10,915 psi) at 28 days were 1.6 and 2.3 times higher, respectively, compared to abrasion-resistant steel. However, when the depths of abrasion were compared as a percentage of the thickness for typical panels of each material, it was concluded that the durability of the high-strength concrete was equivalent to that of the abrasion-resistant steel. Guidance on proportioning of high-strength concrete mixtures is given in EM 1110-2-2000. A thorough discussion of high-strength concrete is given in ACI 363R.

(b) Uplift. Uplift forces caused by high-velocity water flowing over the panel surface is a concern in repair of hydraulic structures, particularly stilling basins. The Old River Low Sill Control Structure demonstrated the problem of uplift of steel panels (McDonald 1980). Thirty modules, 7.3 m (24 ft) long and ranging in width from 0.9 to 6.7 m (3 to 22 ft), were prefabricated from 13-mm (1/2-in.)-thick steel plate for the stilling basin repair. After the modules were installed and anchored underwater, the voids between the steel plate and the existing concrete slab were filled with grout. Additional anchors were installed in holes drilled through the grouted modules and into the basin slab to a depth of 0.9 m (3 ft). An underwater inspection of the basin 8 months after the repairs showed that 7 of the 30 modules had lost portions of their steel plate, ranging from 20 to 100 percent of the surface area. A number of anchor bolts were found broken flush with the surface plate or the grout or pulled completely out of the substrate. A second inspection, approximately 2 years after repair, revealed that additional steel plate had been ripped from four of the modules previously damaged, and an additional nine modules had sustained damage. There were only a few remnants of the steel plates when the basin was dewatered 11 years after the repairs. Precast concrete panels, with a minimum thickness of about 100 mm (4 in.), should be less susceptible to these problems. The increased panel stiffness, damping, and bond to the infill concrete will reduce uplift problems.

(c) Anchors. The design of the anchor system should ensure that the prefabricated panels are adequately anchored to the existing concrete to resist the uplift forces and vibrations created by flowing water. Welding of anchor systems as nearly flush with the prefabricated steel plate surface as possible appears more desirable than raised bolted connections. Preformed, recessed holes for anchors were easily incorporated during precasting of the concrete panels used in underwater repairs at Gavins Point Dam (Section 8-5). The ability of the anchor system, including any embedment material, to perform satisfactorily under the exposure conditions, particularly creep and fatigue, should be evaluated during design of the repair.

(d) Joints. Joint details are important design considerations because once a panel fails and is displaced, adjacent panels are more susceptible to failure as a result of their exposed edges. Consequently, each panel should be designed as an individual repair unit and should not rely on adjacent panels for protection. The vulnerability of joints can be reduced by providing stiffened and recessed panel edges. The joint between the existing concrete and the leading edge of the upstream panels should be designed to provide a smooth transition to the repair section. A general design philosophy should be to minimize the number of joints by using the largest panels practical.

(e) Weight. Panels are usually installed with a barge-mounted crane. Therefore, panel weight can be important, depending on the lifting capacity of available cranes. For panels of a given area, the weight of concrete panels in air will be about four times the weight of steel panels.

(f) Panel supports. Panel supports are bottom-installed platforms or seats which may be required to ensure that panels are placed at the desired elevation and are properly aligned. Rail and Haynes (1991) present several concepts for panel supports, some of which include a means to attach and anchor the panel to resist uplift forces. The potential of these concepts should be evaluated based on specific project requirements.

(g) Infill placement. The physical dimensions of the placement area dictate whether portland-cement grout or concrete is used as the infill material. Grout should be used in those applications that require the infill to flow substantial distances in very thin lifts, whereas pumped concrete or preplaced-aggregate concrete should be used to fill larger voids.

d. Inspection. Inspection of the work as it progresses is important, and there are a number of techniques and procedures for underwater inspection described in Section 2-4. The selection of inspection techniques will be affected by type, size, location, and environmental conditions of a particular job, along with technical capabilities and limitations and monetary constraints. Visual inspection is usually the first technique considered for underwater inspection. Low-light video is an alternative to the diver-inspector. A diver with two-way communication can position the video camera as directed by the inspector who views the video from the surface. A remotely operated vehicle (ROV) with a video camera can also be used where water currents and turbulence permit use of an ROV. More sophisticated techniques such as echo sounders, side-scan sonar, radar, laser mapping, and high-resolution acoustic mapping are also available. The high-resolution acoustic mapping system appears to be ideal for inspection of underwater repairs.

e. Support personnel/equipment. A qualified diving team is required for underwater repairs. Personnel should be skilled in the operation of construction and inspection equipment to be used during the repair. All personnel should have a thorough understanding of proper working procedures to minimize the risk of an accident. Personnel safety should be given a high priority.

8-7. Geomembrane Applications

Many of the Corps' structures exhibit significant concrete cracking which allows water intrusion into or through the structure. These cracks are the result of a variety of phenomena, including restrained concrete shrinkage, thermal gradients, cycles of freezing and thawing, alkali-aggregate reaction, and differential settlement of the foundation. Water leakage through hydraulic structures can also result from poor concrete consolidation during construction, improperly prepared lift or construction joints, and waterstop failures. When leakage rates become unacceptable, repairs are made. Conventional repair methods generally consist of localized sealing of cracks and defective joints by cementitious and chemical grouting, epoxy injection, or surface treatments. Even though localized sealing of leaking cracks and defective joints with conventional methods has been successful in some applications, in many cases some type of overall repair is still required after a few years. Consequently, the potential for geomembranes in such repairs was evaluated as part of the REMR Research Program.

a. Background. Various configurations of geomembranes have been used as impervious synthetic barriers in dams for more than 30 years. Generally, membranes are placed within an embankment or rockfill dam as part of the impervious core or at the upstream face of embankment, rockfill, and concrete gravity dams.

In recent years, geomembranes have been increasingly used for seepage control in a variety of civil engineering structures, including canals, reservoirs, storage basins, dams, and tunnels. Geomembranes have also been used successfully to resurface the upstream face of a number of old concrete and masonry dams, particularly in Europe.

b. Definition. Consistent with the International Commission on Large Dams (ICOLD 1991) the term "geomembrane" is used herein for polymeric membranes which constitute a flexible, watertight material with a thickness of one-half to a few millimeters. A wide range of polymers, including plastics, elastomers, and blends of polymers, are used to manufacture geomembranes.

c. Fabrication. Since the existing concrete or masonry surfaces to receive the geomembrane are usually rough, a geotextile is often used in conjunction with the geomembrane to provide protection against puncturing. For example, the geocomposite SIBELON CNT consists of a nonwoven, needle-punched polyester geotextile which is bound to one side of the flexible PVC geomembrane by heating during the extrusion process. The geotextile is also designed to function as a drain to evacuate any water between the concrete and the geomembrane. The PVC and geotextile layers are 2.5 and 1.5 mm (0.1 and 0.6 in.) thick, respectively, and weigh 3,250 and 500 g/sq m (6 and 0.9 lb/sq yd), respectively. The geocomposite is manufactured in rolls with a minimum width of 2.05 m (6.7 ft) and lengths dependent upon the specific application. In dam applications, each roll is long enough to cover the height of the upstream face where it is to be installed, thus avoiding horizontal welds.

d. Installation. In early applications, geomembranes were attached directly to the upstream face of concrete dams with nails or adhesives. In more recent applications, geocomposites have been installed with stainless steel profiles anchored to the face of the dam. This system (SIBELON SYSTEMS DAMS/CSE) consists of two vertical U-shaped sections fabricated to fit one inside the other to form a continuous rib (Figure 8-58). In addition to a uniform, continuous anchorage of the geocomposite, the system also allows the geomembrane to be pretensioned, thus eliminating the problem of sagging caused by the weight of the geomembrane.

(1) Profiles are not installed near vertical monolith joints to allow the elastic geocomposite to accommodate longitudinal joint movements. The vertical anchorage and tensioning profiles are fabricated in 1.4-m (4.6-ft) lengths with slots for the threaded rods to allow relative vertical displacements between adjacent monoliths.

(2) Any water which might collect behind the geocomposite is conveyed, at atmospheric pressure, along the profiles to the heel of the dam where it can be collected in drainage pipes. A high-density polyethylene geonet, 4 mm (0.016 in.) thick with a diamond-shaped mesh, can be installed behind the geocomposite to increase the drainage capacity of the system. The perimeter of the membrane system is sealed against the concrete face with

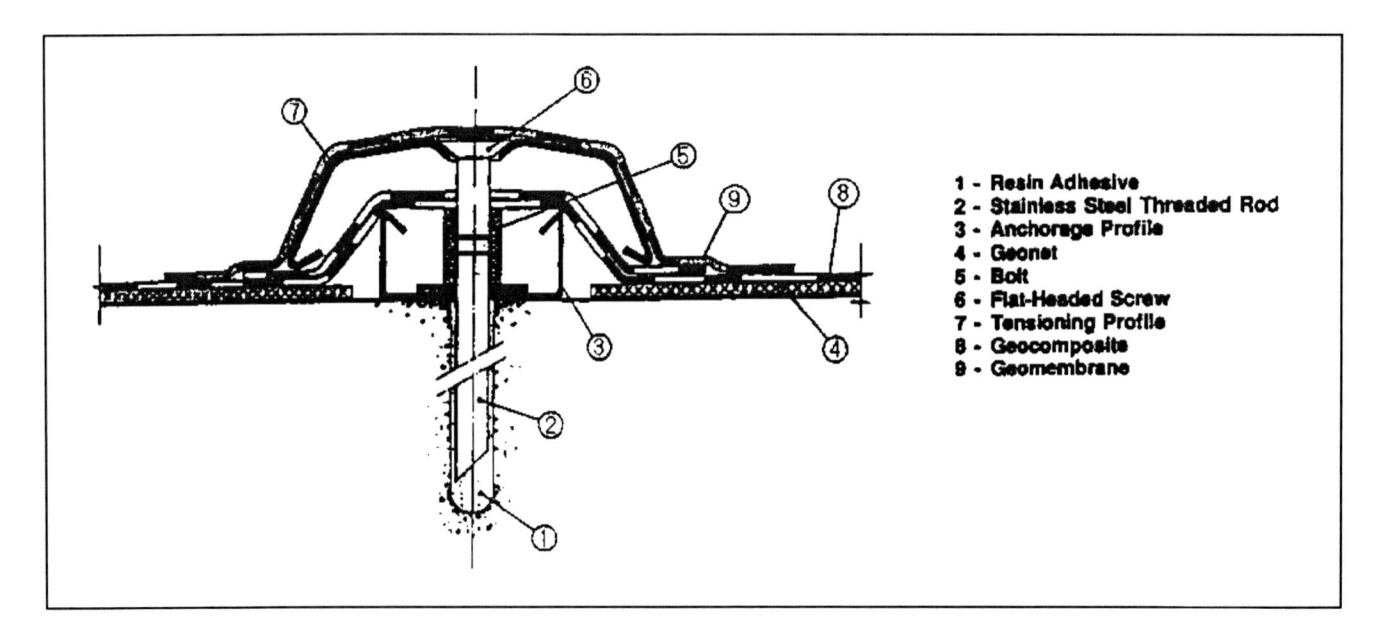

1 - Resin Adhesive
2 - Stainless Steel Threaded Rod
3 - Anchorage Profile
4 - Geonet
5 - Bolt
6 - Flat-Headed Screw
7 - Tensioning Profile
8 - Geocomposite
9 - Geomembrane

Figure 8-58. Detail of vertical anchorage and drainage profile

stainless steel profiles to prevent intrusion of reservoir water. A flat profile is used for the horizontal seal at the crest of the dam and C-shaped profiles (Figure 8-59) are used along the heel of the dam and the abutments.

  e. Applications. Geomembranes have been used to rehabilitate concrete, masonry and rockfill, gravity dams and concrete arch dams including multiple and double curvature arches. Geomembranes have also been used to rehabilitate reservoirs and canals and to provide a water-retention barrier on the upstream face of new dams constructed with roller compacted concrete.

  (1) Lake Baitone Dam. The first Italian application of a geomembrane for rehabilitation purposes was at the 37-m (121-ft)-high Lake Baitone Dam (Cazzuffi 1987). The upstream face of the stone masonry and cement mortar structure, completed in 1930, was lined with a series of vertical, semicircular concrete arches with an internal diameter of 1.7 m (5.6 ft). Deterioration of the concrete surfaces required rehabilitation in 1970 with a 2-mm (.08-in.)-thick polyisobutylene geomembrane applied directly on the arches with an adhesive, without any external protection. After more than 20 years in service, the geomembrane exhibited good adhesion, and its

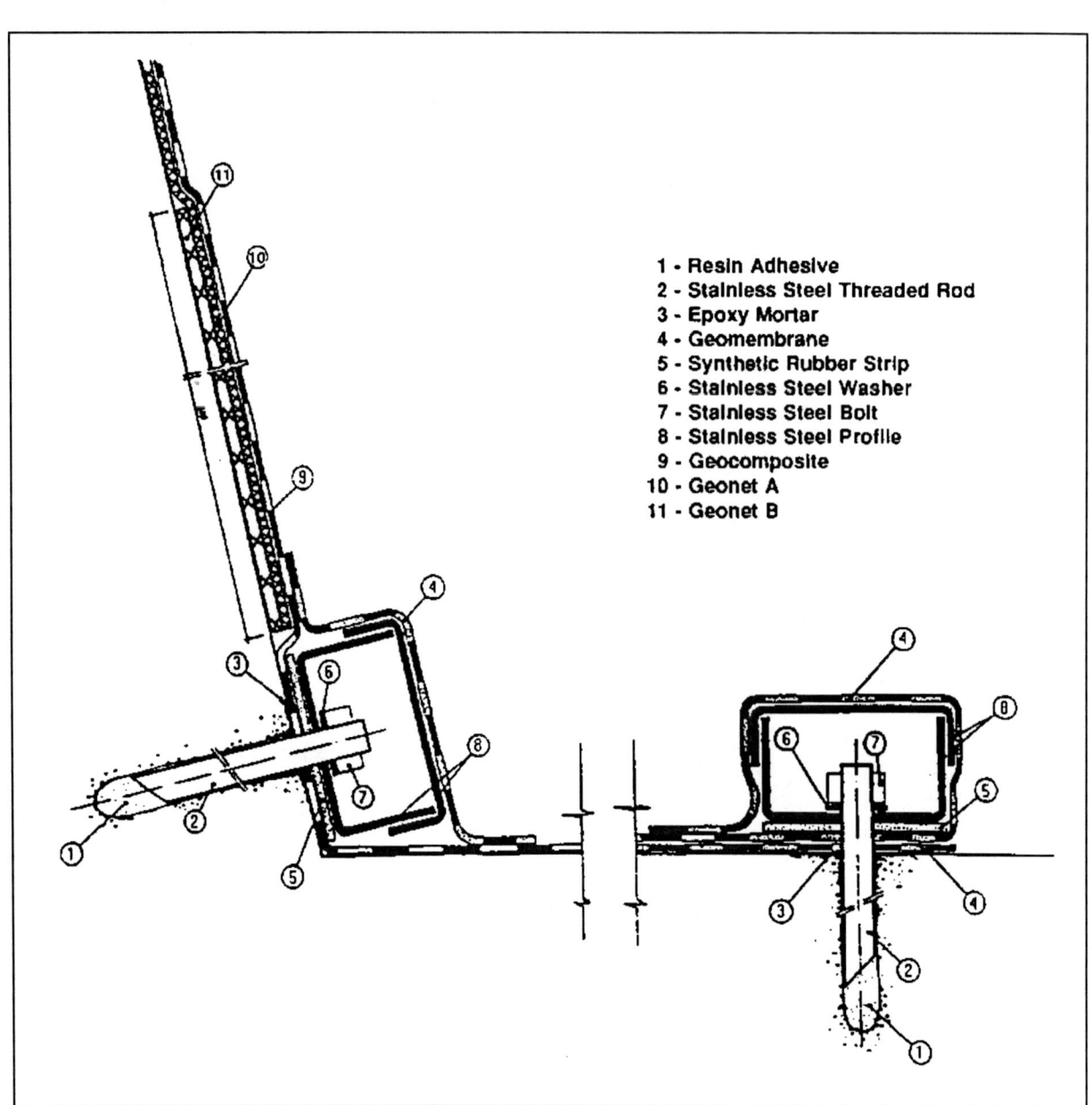

1 - Resin Adhesive
2 - Stainless Steel Threaded Rod
3 - Epoxy Mortar
4 - Geomembrane
5 - Synthetic Rubber Strip
6 - Stainless Steel Washer
7 - Stainless Steel Bolt
8 - Stainless Steel Profile
9 - Geocomposite
10 - Geonet A
11 - Geonet B

Figure 8-59. Detail of sealing profiles

performance was considered to be satisfactory, although there had been some damage caused by heavy ice formation which was quickly and easily repaired. However, when similar repairs were later made on concrete gravity dams, the results were clearly negative (Monari and Scuero 1991). This difference in performance was attributed to the extensive network of cracks in the thin arches which provided for natural drainage of vapor pressure in contrast to the limited drainage associated with the thicker gravity sections with minimal cracking. This experience and subsequent laboratory tests led to the conclusion that geomembrane repair systems should be mechanically anchored and must permit drainage of any water which might be present behind the geomembrane.

(2) Lake Nero Dam. This 40-m (131-ft)-high concrete gravity dam with a crest length of 146 m (479 ft) was completed in 1929 near Bergamo, Italy. Over the years, various repairs, including grouting of the bedrock and shotcreting of the upstream face were conducted. However, these attempts to eliminate leakage through the dam and foundation and deterioration of the concrete caused by aggressive water and cycles of freezing and thawing proved to be temporary or inadequate.

(a) In 1980, a geocomposite was installed on the upstream face of the dam and anchored with steel profiles (Monari 1984). Self-hoisting platforms secured to the dam's crest were used in the installation (Figure 8-60) which required 90 days to complete. The geocomposite was unrolled from the top of the dam down to the base. Adjacent sheets were overlapped and the vertical joints were welded prior to horizontal prestressing. The lower ends of the sheets were anchored to the base of the dam with metal plates.

(b) Installation of the geocomposite reduced leakage from 50 l/sec to 0.27 l/sec with only 14 percent of the leakage through the geocomposite. After 10 years in service, the geocomposite had required no maintenance nor had the lining lost no of its original efficiency (Monari and Scuero 1991).

(3) Cignana Dam. This 58-m (190 ft)-high concrete gravity dam with a crest length of 402 m (1,319 ft) was completed in 1928. Located at an elevation of 2,173 m (7,129 ft) above sea level, the concrete exhibited evidence of considerable freeze-thaw degradation. Also, the dam had major leakage problems despite extensive maintenance efforts, including application of paint-on resin membranes. A 2.5-mm (0.1-in.)-thick geocomposite was

used to rehabilitate the dam in 1987. The repair was similar to that of Lake Nero Dam except that the vertical steel profiles were embedded in a layer of reinforced shotcrete used to resurface the upstream face of the dam. After removal of deteriorated concrete, a layer of shotcrete was applied, vertical profiles were installed, and the surface between adjacent profiles was made level by filling with a second layer of shotcrete (Figure 8-61).

(4) Pracana Dam. This 65-m (213-ft)-high concrete buttress dam with a crest length of 240 m (787 ft) is located on the Ocreza River in Portugal. Cracking occurred soon after construction, causing leakage at the downstream face. An attempt to reduce leakage by grouting the cracks did not result in a durable solution to the problem; therefore, a comprehensive rehabilitation was initiated in 1992. Application of a geocomposite to the upstream face of the dam was a major component of the rehabilitation.

(a) A high-density polyethylene geonet was attached to the upstream face to increase the capacity of the geomembrane system to drain moisture from the concrete and any leakage through the geocomposite. An additional geonet was installed for a height of about 1 m (3 ft) above the heel of the dam to assist in conveying the water to drainage pipes. Ten near-horizontal holes were drilled from the heel of the dam to the downstream face to accommodate the drainage pipes. The geomembrane system was divided into six independent compartments from which water is piped to the downstream face for volumetric measurements. This design provides improved monitoring of the drainage behind the geomembrane.

(b) Approximately 8,000 sq m (9,600 sq yd) of geocomposite were installed with steel profiles shown in Figures 8-58 and 8-59. Vertical joints between rolls of the geocomposite and geomembrane strips covering the profiles (Figure 8-62) were welded to provide a continuous water barrier on the upstream face of the dam. A hot-air jet was used to melt the plastic, and the two surfaces were then welded together by hand pressure on a small roller.

(c) A wire system was installed behind the geocomposite to aid in future monitoring and maintenance of the geomembrane. A sensor sliding on the surface of the geomembrane can detect any anomalies in the electrical field thus locating any discontinuities in the geomembrane.

a. Prior to rehabilitation

b. During rehabilitation

Figure 8-60. Upstream face of Lake Nero Dam

Figure 8-61. Steel profiles embedded in shotcrete, Cignana Dam

Figure 8-62. Installation of geocomposite, Pracana Dam

f. Summary.

(1) Geomembranes and geocomposites have been installed on the upstream face of more than 20 old concrete dams during the past 23 years. The success of these systems in controlling leakage and arresting concrete deterioration and the demonstrated durability of these materials are such that these systems are considered competitive with other repair alternatives.

(2) With a few exceptions, geomembrane installations to date have been accomplished in a dry environment by dewatering the structure on which the geomembrane is to be installed. Dewatering, however, can be extremely expensive and in many cases may not be possible because of project constraints.

(3) A geomembrane system that could be installed underwater would have significantly increased potential in repair of hydraulic structures. Consequently, research has been initiated to develop a procedure for underwater installation of geomembrane repair systems. The objectives of this REMR research are to (a) develop concepts for geomembrane systems that can be installed underwater to minimize or eliminate water intrusion through cracked or deteriorated concrete and defective joints, and (b) demonstrate the constructibility of selected concepts on dams and intake towers.

8-8. RCC Applications

The primary applications of roller-compacted concrete (RCC) within the Corps of Engineers have been in new construction of dams and pavement. Meanwhile, RCC has been so successful for repair of non-Corps dams that the number of dam repair projects now exceeds the number of new RCC dams. The primary advantages of RCC are low cost (25 to 50 percent less than conventionally placed concrete) and rapid construction. Guidance on the use of RCC in civil works structures is given in EM 1110-2-2006.

RCC has been used to strengthen and improve the stability of existing dams, to repair damaged overflow structures, to protect embankment dams during overtopping, and to raise the crest on existing dams. Generally, there are three basic types of dam repairs with RCC: (a) an RCC buttress is placed against the downstream face of an existing dam to improve stability; (b) RCC is used to armor earth and rockfill dams to increase spillway capacity and provide erosion protection should the dam be overtopped; and (c) a combination of the first two, a buttress section and overtopping protection. In addition, RCC has been used to replace the floor in a navigation lock chamber, to help prevent erosion downstream of a floodway sill, and to construct emergency spillways. Selected applications of RCC in repair of a variety of structures are summarized in the following case histories.

a. Ocoee No. 2 Dam. This 9.1-m (30-ft)-high, rock-filled timber crib dam was completed in 1913, near Benton, TN. A riprap berm was placed on the downstream side of the dam in 1980 to improve the stability of the dam. However, flash floods during reconstruction resulted in four washouts of the rock and forced the agency to adopt another method for rehabilitation. The new solution was the first application of RCC to provide increased erosion resistance during overtopping (Hansen 1989). Approximately 3,440 cu m (4,500 cu yd) of RCC was placed in stairsteps on the downstream face of the dam (Figure 8-63). The dam has been subjected to periodic overtopping and, where the RCC was well compacted, it remains undamaged by water flow and weathering (Figure 8-64).

b. New Cumberland Lock. RCC was used to pave the floor of the lock chamber between the upstream emergency bulkhead gate sill and the upstream miter gate sill (Anderson 1984). A drop pipe was used to transfer the RCC from the top of the lockwall to the floor where it was moved by a front end loader, leveled with a dozer, and compacted with a vibratory roller (Figure 8-65). The RCC was placed in 0.3-m (1-ft) layers to a maximum thickness of 1.5 m (5 ft). The nominal maximum size aggregate (NMSA) was 25.4 mm (1 in.), and the cement

content of the mixture was 208 kg per cu m (350 lb per cu yd). A total of 1,645 cu m (2,152 cu yd) of RCC was placed at a cost of $95 per cu m ($73 per cu yd).

c. Chena Project. RCC was used by the U.S. Army Engineer District, Alaska, in lieu of riprap for erosion protection downstream of a floodway sill near Fairbanks. RCC was chosen, even though it was slightly more expensive than riprap, because riprap was in short supply in the area and RCC costs much less than conventionally placed concrete in Alaska (Anderson 1984). The RCC was placed in a section 1.5 m (5 ft) thick, about 12.2 m (40 ft) wide, and 610 m (2,000 ft) long. A total of 12,776 cu m (16,700 cu yd) was placed in 14 days. The cost of the RCC, including cement, was $88 per cu m ($67 per cu yd) compared to $220 per cu m ($168 per cu yd) for conventionally placed concrete on the same project.

d. Kerrville Ponding Dam. The dam consists of a compacted clay embankment 6.4 m (21 ft) high and 182 m (598 ft) long with a 203-mm (8-in.)-thick reinforced concrete cap. The dam, located on the Guadalupe River at Kerrville, TX, was completed in 1980. A service spillway to handle normal flows was created by lowering a 60-m (198-ft) section of the dam 0.3 m (1 ft). The entire dam was designed to be overtopped during flood condition; however, the dam sustained some damage when overtopped in 1981 and 1982. The worst overtopping, 3 m (10 ft) on New Year's Eve, 1984, damaged both the embankment and concrete cap on the downstream slope. The majority of the damage was to the service spillway where about 40 percent of the concrete facing was lost along with a significant portion of the embankment.

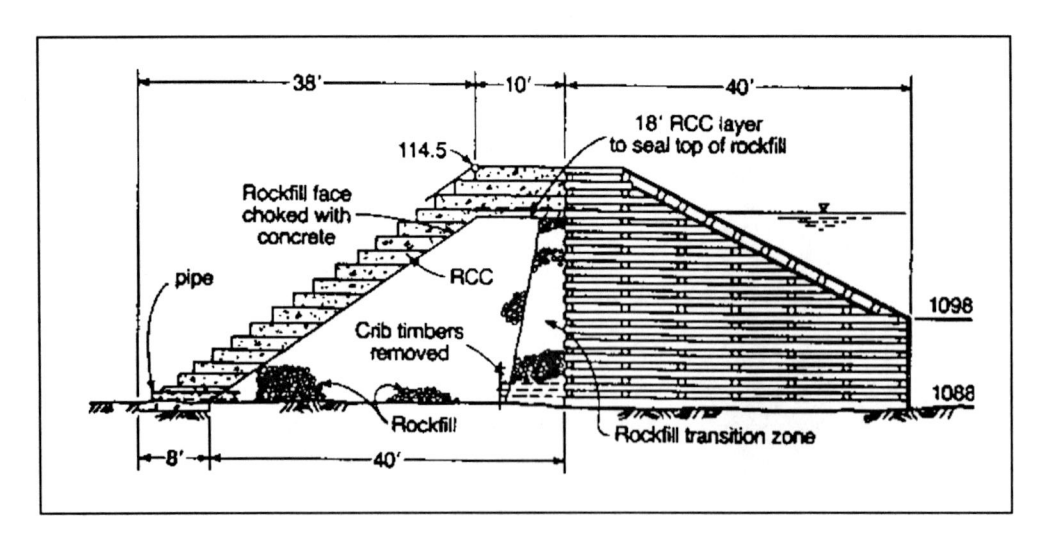

Figure 8-63. Section through modified dam, Ocoee Dam No. 2 (Hansen 1989)

Figure 8-64. Condition of RCC following periodic overtopping, Ocoee No.2 Dam

Figure 8-65. RCC placement, New Cumberland Lock (Anderson 1984)

(1) After several repair alternatives were evaluated, the fastest and most practical solution was to construct an RCC section immediately downstream of the dam (Figure 8-66). The downstream portion of the embankment was removed; the undamaged upstream portion acted as a cofferdam during construction of the RCC section. An 89-mm (3.5-in.) maximum size pit-run aggregate with 10 percent cement by dry weight was used at the base of the dam and in the last five 0.3-m (1-ft)-thick lifts. Cement content was reduced to 5 percent for the middle of the gravity section. The average compressive strength of the richer RCC mixture was 14.5 MPa (2,100 psi) at 28 days. Placement of the 17,140 cu m (22,420 cu yd) of RCC began in late June 1985 and was completed in about 3 months.

(2) About 1 month after completion of RCC placement, a 50-year flood overtopped the structure by 4.4 m (14.4 ft). Water flowed over the entire dam for about 5 days with a maximum flow of 3,540 cu m per sec (125,000 cu ft per sec). An inspection following the flood revealed only minor erosion of uncompacted material at the surface of the weir. Less than 2 years later, a 100-year flood overtopped the structure by 4.9 m (16.2 ft) with a maximum flow of 4,590 cu m per sec (162,000 cu ft per sec). Once again, the RCC performance was outstanding with only minor surface spalling.

e. Boney Falls Dam. The Boney Falls hydroelectric project is located on the Escanaba River about 40 km (25 mi) upstream of Lake Michigan. The dam, which was built during 1920-1921, consists of a 91-m (300-ft)-long right-embankment dam, a nonoverflow gravity dam, a three-unit powerhouse, a gated spillway with six tainter gates, a 61-m (200-ft)-long ungated spillway, and a 1,707-m (5,600-ft)-long left-embankment dam. Normal operating head on the power plant is approximately 15 m (50 ft).

(1) In 1986, it was determined that the original spillways were not adequate to pass the Probable Maximum Flood (PMF) under present-day dam safety criteria. After studying several alternatives, it was concluded that RCC overtopping protection for 305 m (1,000 ft) of the left embankment would be the least costly plan for increasing spillway capacity (Marold 1992). This plan was eventually modified so that an RCC gravity section would be placed immediately behind an existing concrete core wall in the embankment. This modification reduced the length of the RCC spillway section to 152 m (500 ft). The shorter length required the crest of the RCC gravity section to be lower to pass the same flow as the longer paved section. Therefore, the crest of the new RCC gravity section was set at 0.3 m (1 ft) below normal pool and a 1.2-m (4-ft)-thick fuse plug of erodible earthfill was placed on top of the RCC (Figure 8-67).

(2) The 3,710 cu m (4,850 cu yd) of RCC used in the spillway section was placed in 8 days at a cost of $76/cu m ($58/cu yd). When placement was completed, full-depth cores were taken from the spillway section for testing. Core recovery was 100 percent with 97-percent solid concrete and excellent bond between lifts. The average compressive strength of field-cast cylinders was just over 27 MPa (4,000 psi) at 28 days (PCA 1990).

f. Camp Dyer Diversion Dam. The dam, which was completed in 1926, is located on the Auga Fria River

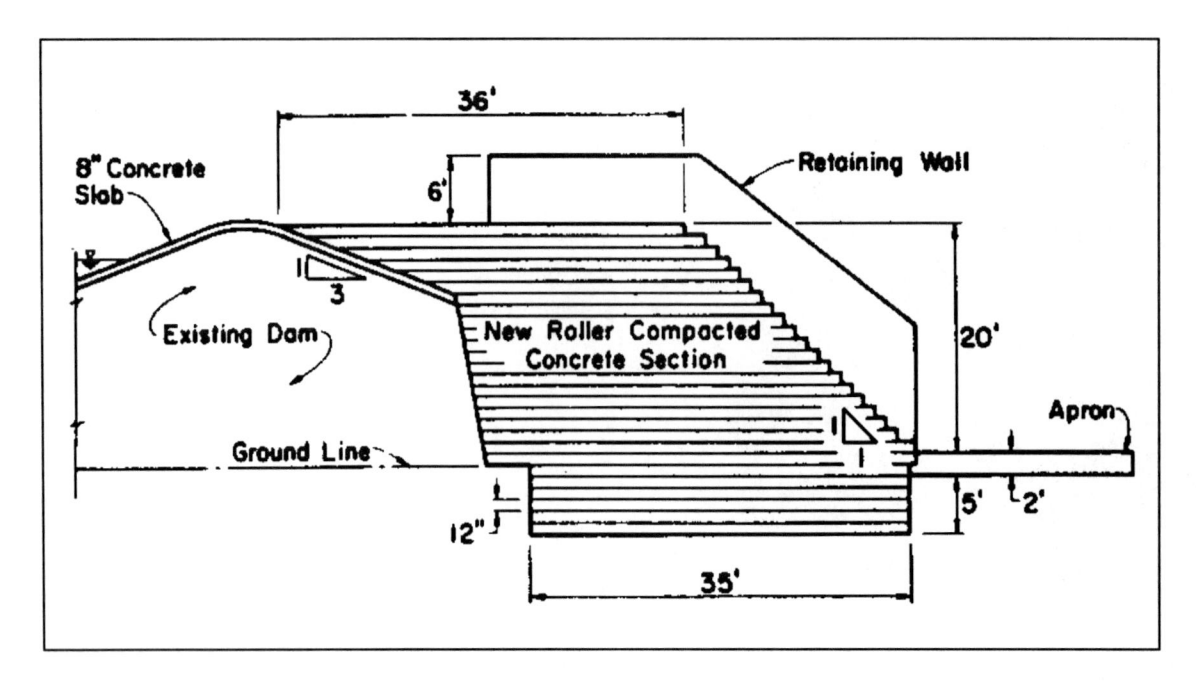

Figure 8-66. RCC gravity section, Kerrville Ponding Dam (Hansen and France 1986)

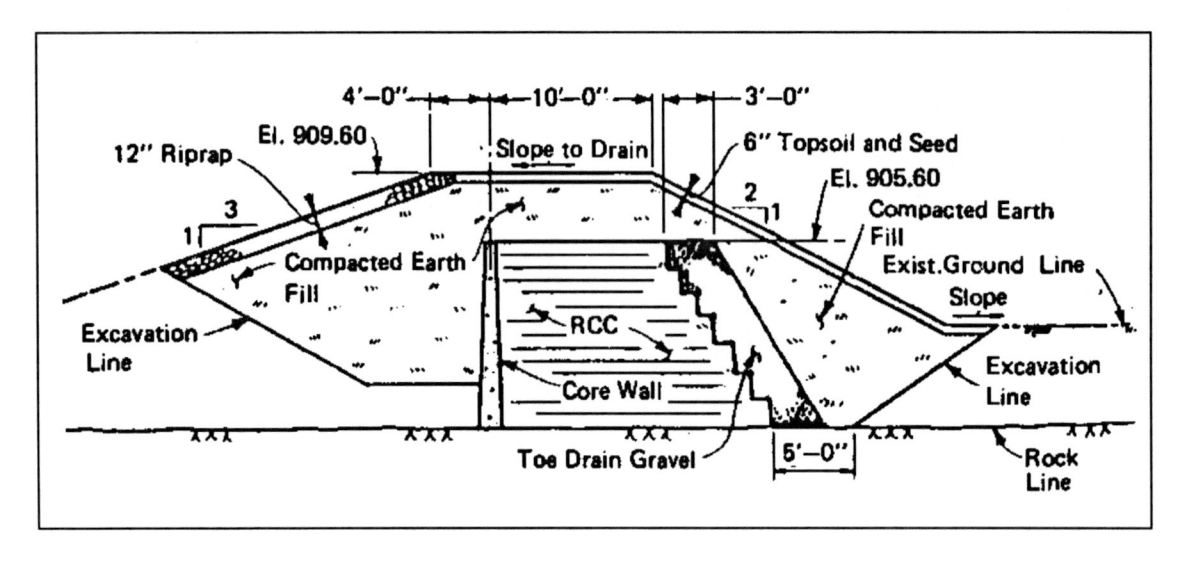

Figure 8-67. RCC spillway section, Boney Falls Dam (Marold 1992)

about 56 km (35 mi) northwest of Phoenix, AZ, and about 1.6 km (1 mi) downstream from Waddell Dam. The original masonry and concrete gravity structure had a crest length of 187 m (613 ft) and a maximum structural height of 23 m (75 ft). A smaller concrete gravity dike west of the dam had an 80-m (263-ft) crest length and a 7.6-m (25 ft) maximum structural height. Construction of the New Waddell Dam midway between the Waddell and the Camp Dyer Diversion Dams significantly reduced the storage capacity of the lower lake. Consequently, the height of the Camp Dyer Diversion Dam was raised

1.2 m (3.9 ft) to maintain the lake's original storage capacity, and the dam was modified to meet current criteria for static and dynamic stability of concrete gravity dams (Hepler 1992).

(1) To increase the dead load and the sliding resistance of the structure, an RCC buttress was designed for the downstream face of the existing dam. RCC was selected over conventional concrete because of its relative economy and ease of construction. In the original design, the buttress had a nominal width of 4.6 m (15 ft) and a

downstream slope of 8H:1V. However, in the final design, the width was increased to 6.1 m (20 ft) to accommodate two lanes of construction traffic on the RCC lifts. The RCC buttresses were capped with a conventional reinforced-concrete apron and ogee overflow crest (Figure 8-68).

(2) Approximately 11,800 cu m (15,400 cu yd) of RCC was used to construct the dam and dike buttresses. Total cost of the project, including the RCC buttresses and associated work, was about $3 million.

g. Gibraltar Dam. The dam is located on the Santa Ynez River north of Santa Barbara, CA. The original dam was completed in 1922 and raised in 1948 to provide storage for the city's municipal water supply. It is a constant radius, concrete arch dam with a maximum height of 59 m (195 ft) and a crest length of 183 m (600 ft). The thickness of the arch varies from 2.1 m

(7 ft) at the crest to approximately 20 m (65 ft) at the base.

(1) The results of a 1983 safety evaluation indicated that the dam did not meet seismic safety standards. Therefore, an RCC buttress was constructed against the downstream face of the dam to alter the dynamic response characteristics of the dam and thus reduce the stresses induced during an earthquake. The addition of the gravity buttress effectively converted the existing arch into a curved gravity dam as shown in Figure 8-69 (Wong et al. 1992).

(2) The RCC method of strengthening was selected after three alternate methods were evaluated in terms of design, environment, construction, and cost. The other methods considered were placing a blanket of reinforced concrete or shotcrete over both faces of the dam, and constructing a rockfill buttress against the downstream

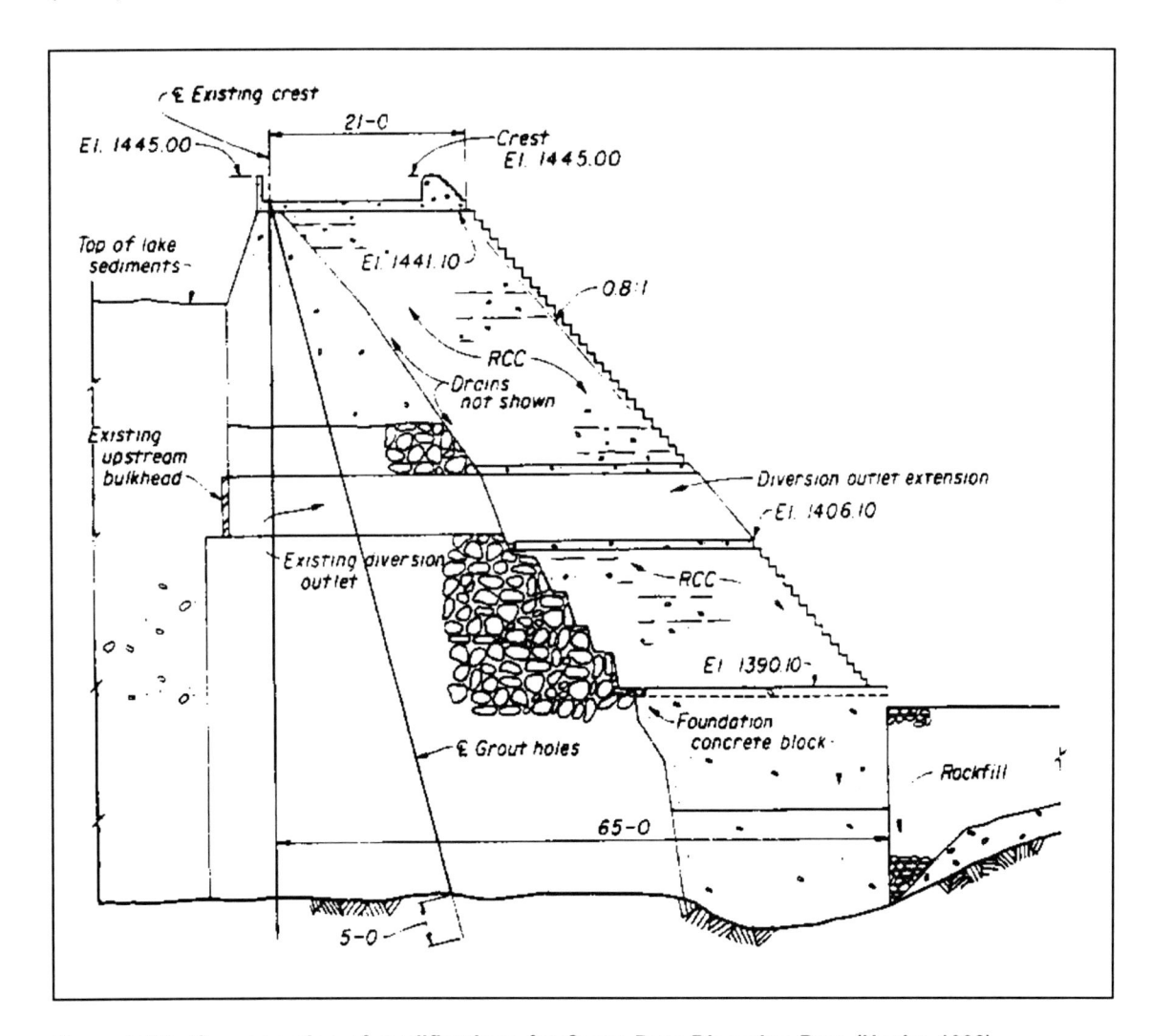

Figure 8-68. Cross section of modifications for Camp Dyer Diversion Dam (Hepler 1992)

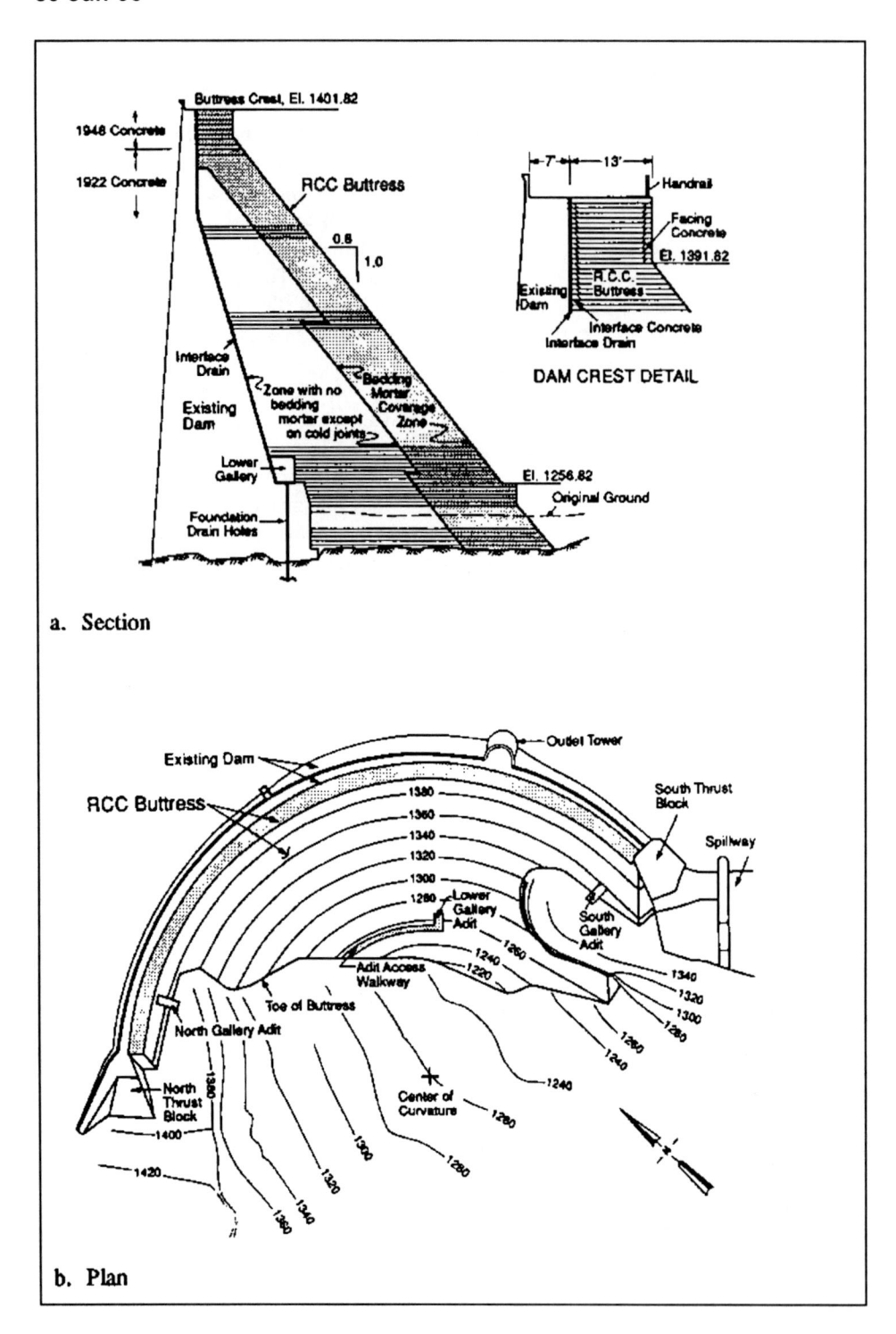

a. Section

b. Plan

Figure 8-69. Section and plan of modified Gibraltar Dam (Wong et al 1992)

face. RCC was chosen because of the speed of placement that would translate to lower construction costs. The constricted wilderness site also made swift placement an advantage, and potential for environmental impact by other methods was a third factor in the choice.

(3) All onsite construction, including placement of 71,870 cu m (94,000 cu yd) of RCC, was completed in 1 year at a total cost of $8.18 million.

h. Littlerock Dam. This multiple-arch, reinforced-concrete dam is located near Palmdale, CA, about 2.4 km (1.5 mi) south of the San Andreas fault. The historically significant dam, completed in 1924, has a maximum height of 53 m (175 ft) and a crest length of 219 m (720 ft). The mutipile-arch structure that comprises the main section of the dam consists of 24 arch bays with buttresses at 7.3-m (24-ft) centers. Results of stability and stress analyses showed that the dam did not meet required seismic safety criteria, principally because of a lack of lateral stability, a deficiency inherent in multipile-arch dams (Wong, Forrest, and Lo 1993).

(1) To satisfy preservation requirements and to provide adequate seismic stability, engineers designed a

rehabilitation program that included construction of an RCC gravity section between and around the downstream portions of the existing buttresses (Figure 8-70). Also, steel fiber-reinforced shotcrete with silica fume was used to resurface and stiffen the arches of the existing dam. RCC placement began in November 1993, and approximately 70,500 cu m (93,000 cu yd) was placed in just over 2 months. The RCC essentially converted the multiple-arch dam into a gravity structure. In addition, the crest of the dam was raised 3.7 m (12 ft) with conventional concrete to increase water storage capacity. The cost of the RCC buttress repair was $12.8 million compared to an estimated cost of $22.5 million for filling the arch bays with mass concrete.

(2) When the Northridge earthquake shook southern California in January 1994, the RCC placement was within 5 m (16 ft) of the crest of the existing dam. Although there was widespread damage to highway bridges within 56 km (35 mi) of the project, the quake had no noticeable impact on the dam.

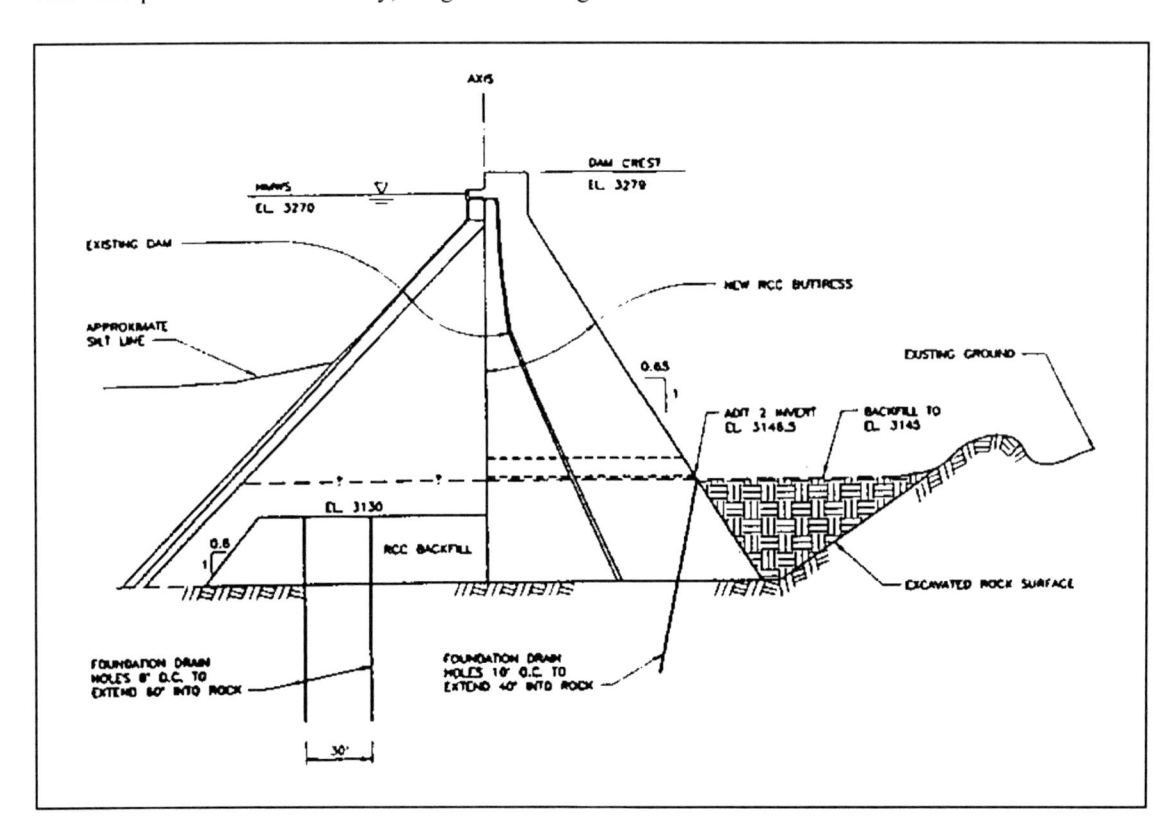

Figure 8-70. Littlerock Dam modification (Bischoff and Obermeyer 1993)

Chapter 9
Concrete Investigation Reports

9-1. General

When required by a major rehabilitation project, a concrete materials design memorandum, in the form of a separate report or a part of The Rehabilitation Evaluation Report, will be prepared. The need for repairs can vary from such minor problems as shrinkage cracks and pop-outs to major damage resulting from severe cavitation or abrasion or from structural failure. Therefore, the scope of the investigation will also vary, depending on the criticality of the project. For minor (noncritical) repair of concrete structures in moderate climates, the investigation may be limited to determining the availability of satisfactory repair materials. Examples of such repairs include patching small spalls and sealing dormant cracks in southern states. If the environment of the project is known to be deleterious to concrete, such as cycles of freezing and thawing, sulfate attack, or acid attack, the investigation must address the measures to be taken to mitigate deterioration, regardless of the quantity of repair materials involved. A more detailed investigation will normally be required for rehabilitation of structures such as locks, dams, large pumping stations, and power plants.

9-2. Concrete Materials Design Memorandum

In general, a design memorandum will include information as required in Appendix C of EM 1110-2-2000. In addition, for all repair or rehabilitation projects other than minor repairs, the design memorandum will also include the following:

(1) Description of the existing conditions and types of deterioration.

(2) Detailed discussion of the causes of deterioration.

(3) Proposed method(s) of removal of damaged and deteriorated concrete.

(4) Proposed repair materials and detailed repair procedures.

If epoxy or other polymer materials are to be used in the repair, detailed discussion of the materials and justifications for using such materials should also be given. Chapter 4 provides more detailed planning and design considerations. Although the requirements for a materials design memorandum may appear to be unnecessary detail for a project involving only a small quantity of repair materials, experience has repeatedly demonstrated that no step in a repair operation can be omitted or carelessly performed without detriment to the serviceability of the work. Inadequate investigation into the causes of the problem and inadequate workmanship, procedures, or materials will result in inferior repairs which ultimately will fail.

9-3. Concrete Repair Report

At the conclusion of any unique concrete repair project, an as-built concrete repair report should be prepared. Unique repair projects are those on which unusual concrete repair materials and techniques are used. The concrete repair report should include discussions of problems encountered in each phase of the repair work, effectiveness of the repair materials and techniques, and results of the follow-up evaluation program, if available.

Appendix A
References

A-1. Required Publications

TM 5-822-6/AFM 88-7, Chapter 1
Rapid Pavements for Roads, Streets, Walks, and Open Storage Areas

TM 5-822-9/AFM 88-6, Chapter 10
Repair of Rigid Pavements Using Epoxy-Resin Grouts, Mortars, and Concrete

EP 1110-1-10
Borehole Viewing Systems

EM 385-1-1
Safety and Health Requirements Manual

EM 1110-1-3500
Chemical Grouting

EM 1110-2-2000
Standard Practice for Concrete

EM 1110-2-2005
Standard Practice for Shotcrete

EM 1110-2-2006
Roller-Compacted Concrete

EM 1110-2-2102
Waterstops and Other Joint Materials

EM 1110-2-3506
Grouting Technology

EM 1110-2-4300
Instrumentation for Concrete Structures CH1

U.S. Army Engineer Waterways Experiment Station 1949
U.S. Army Engineer Waterways Experiment Station. 1949 (Aug). *Handbook for Concrete and Cement*, with quarterly supplements (all CRD-C designations), Vicksburg, MS. Note: Use latest edition of all designations.

U.S. Army Engineer Waterways Experiment Station 1985
U.S. Army Engineer Waterways Experiment Station. 1985. *The REMR Notebook*, with periodic supplements, Vicksburg, MS, including:

a. "Video Systems for Underwater Inspection of Structures," REMR Technical Note CS-ES-2.6

b. "Underwater Cameras for Inspection of Structures in Turbid Water," REMR Technical Note CS-ES-3.2

c. "Removal and Prevention of Efflorescence on Concrete and Masonry Building Surfaces," REMR Technical Note CS-MR-4.3

d. "Cleaning Concrete Surfaces," REMR Technical Note CS-MR-4.4

e. "General Information of Polymer Materials," REMR Technical Note CS-MR-7.1

f. "Antiwashout Admixtures for Underwater Concrete," REMR Technical Note CS-MR-7.2

g. "Rapid-Hardening Cements and Patching Materials," REMR Technical Note CS-MR-7.3

h. "Handling and Disposal of Construction Residue," REMR Technical Note EI-M-1.2

Handbooks and reports published by the Waterways Experiment Station may be obtained from: U.S. Army Engineer Waterways Experiment Station, 3909 Halls Ferry Road, Vicksburg, MS 39180-6199.

American Concrete Institute (Annual)
American Concrete Institute. Annual. *Manual of Concrete Practice*, Five Parts, Detroit, MI, including:

"Cement and Concrete Terminology," ACI 116R

"Guide for Making a Condition Survey of Concrete in Service," ACI 201.lR

"Guide to Durable Concrete," ACI 201.2R

"Mass Concrete," ACI 207.lR

"Effect of Restraint, Volume Change, and Reinforcement on Cracking of Mass Concrete," ACI 207.2R

"Practices for Evaluation of Concrete in Existing Massive Structures for Service Conditions," ACI 207.3R

"Roller Compacted Mass Concrete," ACI 207.5R

"Erosion of Concrete in Hydraulic Structures," ACI 210R

"Corrosion of Metals in Concrete," ACI 222R

"Standard Practice for the Use of Shrinkage-Compensating Concrete," ACI 223

"Causes, Evaluation, and Repair of Cracks in Concrete Structures," ACI 224.1R

"Guide for Concrete Floor and Slab Construction," ACI 302.1R

"Guide for Measuring, Mixing, Transporting, and Placing Concrete," ACI 304R

"Guide for the Use of Preplaced Aggregate Concrete for Structural and Mass Concrete Applications," ACI 304.1R

"Placing Concrete by Pumping Methods," ACI 304.2R

"Hot Weather Concreting," ACI 305R

"Building Code Requirements for Reinforced Concrete," ACI 318

"Guide for the Design and Construction of Concrete Parking Lots," ACI 330R

"Guide to Residential Cast-in-Place Concrete Construction," ACI 332R

"State-of-the-Art Report on Anchorage to Concrete," ACI 355.1R

"State-of-the-Art Report on High-Strength Concrete," ACI 363R

"Guide for Evaluation of Concrete Structures Prior to Rehabilitation," ACI 364.1R

"Use of Epoxy Compounds with Concrete," ACI 503R

"Standard Specification for Bonding Plastic Concrete to Hardened Concrete With a Multi-Component Epoxy Adhesive," ACI 503.2

"Guide for the Selection of Polymer Adhesives with Concrete," ACI 503.5R

"Guide to Sealing Joints in Concrete Structures," ACI 504R

"Guide to Shotcrete," ACI 506R

"A Guide to the Use of Waterproofing, Dampproofing, Protective, and Decorative Barrier Systems for Concrete," ACI 515.lR

"Guide for Specifying, Proportioning, Mixing, Placing, and Finishing Steel," ACI 544.3R

"Guide for the Use of Polymers in Concrete," ACI 548.1R

"State-of-the-Art Report on Polymer-Modified Concrete," ACI 548.3R

"Standard Specification for Latex-Modified Concrete (LMC) Overlays," ACI 548.4

ACI 226 1987
ACI Committee 226. 1987 (Mar-Apr). "Silica Fume in Concrete," *ACI Materials Journal*, Vol 84, No. 2, pp 158-166.

ACI publications may be obtained from: American Concrete Institute, Member/Customer Services Department, Box 19150, Detroit, MI 48219-0150.

American Society for Testing and Materials (Annual)
American Society for Testing and Materials. Annual. *Annual Book of ASTM Standards*, Philadelphia, PA. Note: Use the latest available issue of each ASTM standard.

ASTM Standards and Publications may be obtained from: American Society for Testing and Materials, 1916 Race Street, Philadelphia, PA 19103.

A-2. Related Publications

ABAM Engineers 1987a
ABAM Engineers, Inc. 1987a (Jul). "Design of a Precast Concrete Stay-in-Place Forming System for Lock Wall Rehabilitation," Technical Report REMR-CS-7, U.S. Army Engineer Waterways Experiment Station, Vicksburg, MS.

ABAM Engineers 1987b
ABAM Engineers, Inc. 1987b (Dec). "A Demonstration of the Constructibility of a Precast Concrete Stay-in-Place Forming System for Lock Wall Rehabilitation," Technical Report REMR-CS-14, U.S. Army Engineer Waterways Experiment Station, Vicksburg, MS.

ABAM Engineers 1989
ABAM Engineers, Inc. 1989 (Dec). "Concepts for Installation of the Precast Concrete Stay-in-Place Forming System for Lock Wall Rehabilitation in an Operational Lock," Technical Report REMR-CS-28, U.S. Army Engineer Waterways Experiment Station, Vicksburg, MS.

Ahmad and Haskins 1993
Ahmad, Falih H., and Haskins, Richard. 1993 (Sep). "Use of Ground-Penetrating Radar in Nondestructive Testing for Voids and Cracks in Concrete," The REMR Bulletin, Vol 10, No. 3, pp 11-15.

Alexander 1980
Alexander, A. M. 1980 (Apr). "Development of Procedures for Nondestructive Testing of Concrete Structures; Report 2, Feasibility of Sonic Pulse-Echo Technique," Miscellaneous Paper C-77-11, U.S. Army Engineer Waterways Experiment Station, Vicksburg, MS.

Alexander 1993
Alexander, A. Michel. 1993 (Apr). "Impacts on a Source of Acoustic Pulse-Echo Energy for Nondestructive Testing of Concrete Structures," Technical Report REMR-CS-40, U.S. Army Engineer Waterways Experiment Station, Vicksburg, MS.

Alexander and Thornton 1988
Alexander, A. M., and Thornton, H. T., Jr. 1988. "Developments in Ultrasonic Pitch-Catch and Pulse-Echo for Measurements in Concrete," SP-112, American Concrete Institute, Detroit, MI.

Alongi, Cantor, Kneeter, Alongi 1982
Alongi, A. V., Cantor, T. R., Kneeter, C. P., and Alongi, A., Jr. 1982. "Concrete Evaluation by Radar Theoretical Analysis," Concrete Analysis and Deterioration, Transportation Research Board, Washington, DC.

Anderson 1984
Anderson, Fred A. 1984 (May). "RCC Does More," Concrete International, American Concrete Institute, Vol 6, No. 5, pp 35-37.

Bach and Isen 1968
Bache, H. H., and Isen, J. C. 1968 (Jun). "Model Determination of Concrete Resistance to Popout Formation," Journal of the American Concrete Institute, Proceedings, Vol 65, pp 445-450.

Bean 1988
Bean, Dennis L. 1988 (Apr). "Surface Treatments to Minimize Concrete Deterioration; Report 1, Survey of Field and Laboratory Application and Available Products," Technical Report REMR-CS-17, U.S. Army Engineer Waterways Experiment Station, Vicksburg, MS.

Best and McDonald 1990a
Best, J. Floyd, and McDonald, James E. 1990 (Jan). "Spall Repair of Wet Concrete Surfaces," Technical Report REMR-CS-25, U.S. Army Engineer Waterways Experiment Station, Vicksburg, MS.

Best and McDonald 1990b
Best, J. Floyd, and McDonald, James E. 1990 (Jan). "Evaluation of Polyester Resin, Epoxy, and Cement Grouts for Embedding Reinforcing Steel Bars in Hardened Concrete," Technical Report REMR-CS-23, U.S. Army Engineer Waterways Experiment Station, Vicksburg, MS.

Bischoff and Obermeyer 1993
Bischoff, John A., and Obermeyer, James R. 1993 (Apr). "Design Considerations for Raising Existing Dams for Increased Storage," Geotechnical Practice in Dam Rehabilitation, American Society of Civil Engineers, pp 174-187.

Bryant and Mlakar 1991
Bryant, Larry M., and Mlakar, Paul F. 1991 (Mar). "Predicting Concrete Service Life in Cases of Deterioration Due to Freezing and Thawing," Technical Report REMR-CS-35, U.S. Army Engineer Waterways Experiment Station, Vicksburg, MS.

Busby Associates, Inc. 1987
Busby Associates, Inc. 1987. "Undersea Vehicles Directory," Arlington, TX.

Campbell 1982
Campbell, Roy L., Sr. 1982 (Apr). "A Review of Methods for Concrete Removal," Technical Report SL-82-3, U.S. Army Engineer Waterways Experiment Station, Vicksburg, MS.

Campbell 1994
Campbell, Roy L., Sr. 1994 (Feb). "Overlays on Horizontal Concrete Surfaces: Case Histories," Technical Report REMR-CS-42, U.S. Army Engineer Waterways Experiment Station, Vicksburg, MS.

Carino 1992
Carino, Nicholas J. 1992 (Jan). "Recent Developments in Nondestructive Testing of Concrete," Advances in Concrete Technology, Canada Center for Mineral and Energy Technology, Ottawa, Canada, pp 281-328.

Carter 1993
Carter, Paul. 1993 (Jan). "Developing a Performance-Based Specification for Concrete Sealers on Bridges," *Journal of Protective Coatings & Linings*, pp 36-44.

Cazzuffi 1987
Cazzuffi, D. 1987 (Mar). "The Use of Geomembranes in Italian Dams," *Water Power & Dam Construction*.

Clausner and Pope 1988
Clausner, J. E., and Pope, J. 1988 (Nov.) "Side-Scan Sonar Applications for Evaluating Coastal Structures," Technical Report CERC-88-16, U.S. Army Engineer Waterways Experiment Station, Vicksburg, MS.

Clear and Chollar 1978
Clear, K. C., and Choller, B. H. 1978 (Apr). "Styrene-Butadiene Latex Modifiers for Bridge Peak Overlay Concrete," Report No. FHWA-RD-35, Federal Highway Administration, Washington, DC.

Clifton 1991
Clifton, James R. 1991 (Nov). "Predicting the Remaining Service Life of Concrete," Report NISTIR 4712, National Institute of Standards Technology, Gaithersburg, MD.

Concrete Repair Digest 1993
Concrete Repair Digest. 1993 (Feb/Mar). "Removing Some Common Stains from Concrete."

Dahlquist 1987
Dahlquist, M. S. 1987 (Oct). *REMR Bulletin*, Vol 4, No. 2, U.S. Army Engineer Waterways Experiment Station, Vicksburg, MS.

Davis 1960
Davis, R. E., "Prepakt Method of Concrete Repair," *ACI Journal*, Vol 32, pp 155-1752.

Davis, Jansen, and Neelands 1948
Davis, R. E., Jansen, E. C., and Neelands, W. T. 1948 (Apr). "Restoration of Barker Dam," *ACI Journal*, Vol 19, No. 8, pp 633-688.

Dobrowolski and Scanlon 1984
Dobrowolski, Joseph A., and Scanlon, John M. 1984 (Jun). "How to Avoid Deficiencies in Architectural Concrete Construction," Technical Report SL-84-9, U.S. Army Engineer Waterways Experiment Station, Vicksburg, MS.

Emmons 1993
Emmons, Peter H. 1993. *Concrete Repair and Maintenance Illustrated*, R. S. Means Co., Inc., Kingston, MA, 295 pp.

Emmons and Vaysburd 1995
Emmons, Peter H., and Vaysburd, Alexander M. 1995 (Mar). "Performance Criteria for Concrete Repair Materials, Phase I," Technical Report REMR-CS-47, U.S. Army Engineer Waterways Experiment Station, Vicksburg, MS.

Emmons, Vaysburd, and McDonald 1993
Emmons, Peter H., Vaysburd, Alexander M., and McDonald, James E. 1993 (Sep). "A Rational Approach to Durable Concrete Repairs," *Concrete International*, Vol 15, No. 9, pp 40-45.

Emmons, Vaysburd, and McDonald 1994
Emmons, P. H., Vaysburd, A. M., and McDonald, J. E. 1994 (Mar). "Concrete Repair in The Future Turn Of The Century - Any Problems?," *Concrete International*, Vol 16, No. 3, pp 42-49.

Fenwick 1989
Fenwick, W. B. 1989 (Aug). "Kinzua Dam, Allegheny River, Pennsylvania and New York; Hydraulic Model Investigation," Technical Report HL-89-17, U.S. Army Engineer Waterways Experiment Station, Vicksburg, MS.

Gerwick 1988
Gerwick, B. C. 1988 (Sep). "Review of the State of the Art for Underwater Repair Using Abrasion-Resistant Concrete," Technical Report REMR-CS-19, U.S. Army Engineer Waterways Experiment Station, Vicksburg, MS.

Gurjar and Carter 1987
Gurjar, Suresh, and Carter, Paul. 1987 (Mar). "Alberta Concrete Patch Evaluation Program," Report No. ABTR/RD/RR-87/05, Alberta Transportation & Utilities, Edmonton, Alberta, Canada.

Hacker 1986
Hacker, Kathy. 1986 (Dec). "Precast Panels Speed Rehabilitation of Placer Creek Channel," *The REMR Bulletin*, Vol 3, No. 3, U.S. Army Engineer Waterways Experiment Station, Vicksburg, MS.

Hammonds, Garner, and Smith 1989
Hammons, M. I., Garner, S. B., and Smith, D. M. 1989 (Jun). "Thermal Stress Analysis of Lock Wall, Dashields

Locks, Ohio River," Technical Report SL-89-6, U.S. Army Engineer Waterways Experiment Station, Vicksburg, MS.

Hansen 1989
Hansen, Kenneth D. 1989 (Oct). "Performance of Roller-Compacted Concrete Dam Rehabilitations," *Proceedings from the 6th ASDSO Annual Conference*, Association of State Dam Safety Officials, pp 21-26.

Hansen and France 1986
Hansen, Kenneth D., and France, John W. 1986 (Sep). "RCC: A Dam Rehab Solution Unearthed,"*Civil Engineering*, American Society of Civil Engineers, Vol 56, No. 9, pp 60-63.

Hepler 1992
Hepler, Thomas E. 1992. "RCC Buttress Construction for Camp Dyer Diversion Dam," *Proceedings from the 9th ASDSO Annual Conference*, Association of State Dam Safety Officials, pp 21-26.

Holland 1983
Holland, T. C. 1983 (Sep). "Abrasion-Erosion Evaluation of Concrete Mixtures for Stilling Basin Repairs, Kinzua Dam, Pennsylvania," Miscellaneous Paper SL-83-16, U.S. Army Engineer Waterways Experiment Station, Vicksburg, MS.

Holland 1986
Holland, T. C. 1986 (Sep). "Abrasion-Erosion Evaluation of Concrete Mixtures for Stilling Basin Repairs, Kinzua Dam, Pennsylvania," Miscellaneous Paper SL-86-14, U.S. Army Engineer Waterways Experiment Station, Vicksburg, MS.

Holland and Gutschow 1987
Holland, Terence C., and Gutschow, Richard A. 1987 (Mar). "Erosion Resistance with Silica-Fume Concrete," *Concrete International*, Vol 9, No. 3, pp 32-40.

Holland, Husbands, Buck, and Wong 1980
Holland, T. C., Husbands, T. B., Buck, A. D., and Wong, G. S. 1980 (Dec). "Concrete Deterioration in Spillway Warm-Water Chute, Raystown Dam, Pennsylvania," Miscellaneous Paper SL-80-19, U.S. Army Engineer Waterways Experiment Station, Vicksburg, MS.

Holland, Krysa, Luther, and Liu 1986
Holland, T. C., Krysa, A., Luther, M. D., and Liu, T. C. 1986. "Use of Silica-Fume Concrete to Repair Abrasion-Erosion Damage in the Kinzua Dam Stilling Basin," *Fly Ash, Silica Fume, Slag, and Natural Pozzolans in*

Concrete, SP-91, Vol 2, American Concrete Institute, Detroit, MI.

Holland and Turner 1980
Holland, T. C., and Turner, J. R. 1980 (Sep). "Construction of Tremie Concrete Cutoff Wall, Wolf Creek Dam, Kentucky," Miscellaneous Paper SL-80-10, U.S. Army Engineer Waterways Experiment Station, Vicksburg, MS.

Houghton, Borge, and Paxton 1978
Houghton, D. L., Borge, O. E., and Paxton, J. H. 1978 (Dec). "Cavitation Resistance of Some Special Concretes," *ACI Journal,* Vol 75, No. 12, pp 664-667.

Hulshizer and Desai 1984
Hulshizer, A. J., and Desai, A. J. 1984 (Jun). "Shock Vibration Effects on Freshly Placed Concrete," *ASCE Journal of Construction Engineering and Management,* Vol 110, No. 2, pp 266-285.

Hurd 1989
Hurd, M. K. 1989. "Formwork for Concrete," 5th Edition, SP-4, American Concrete Institute, Detroit, MI.

Husbands and Causey 1990
Husbands, Tony B., and Causey, Fred E. 1990 (Sep). "Surface Treatments to Minimize Concrete Deterioration; Report 2, Laboratory Evaluation of Surface Treatment Materials," Technical Report REMR-CS-17, U.S. Army Engineer Waterways Experiment Station, Vicksburg, MS.

ICOLD 1991
ICOLD. 1991. *Watertight Geomembranes for Dams - State of the Art,* Bulletin 78, International Commission on Large Dams, Paris.

Johnson 1965
Johnson, S. M. 1965. *Deterioration, Maintenance, and Repair of Structures,* McGraw-Hill, New York.

Kahl, Kauschinger, and Perry 1991
Kahl, Thomas W., Kauschinger, Joseph L., and Perry, Edward B. 1991 (Mar). "Plastic Concrete Cutoff Walls for Earth Dams," Technical Report REMR-GT-15, U.S. Army Engineer Waterways Experiment Station, Vicksburg, MS.

Keeney 1987
Keeney, C. A. 1987 (Nov). "Procedures and Devices for Underwater Cleaning of Civil Works Structures," Technical Report REMR-CS-8, U.S. Army Engineer Waterways Experiment Station, Vicksburg, MS.

Khayat 1991
Khayat, K. H. 1991. "Underwater Repair of Concrete Damaged by Abrasion-Erosion," Technical Report REMR-CS-37, U.S. Army Engineer Waterways Experiment Station, Vicksburg, MS.

Kottke 1987
Kottke, Edgar. 1987 (Aug). "Evaluation of Sealers for Concrete Bridge Elements," Alberta Transportation and Utilities, Alberta, Canada.

Krauss 1994
Krauss, P. D. 1994 (Mar). "Repair Materials and Techniques for Concrete Structures in Nuclear Power Plants," ORNL/NRC/LTR-93/28, Oak Ridge National Laboratory, Oak Ridge, TN.

Kucharski and Clausner 1990
Kucharski, W. M., and Clausner, J. E. 1990 (Feb). "Underwater Inspection of Coastal Structures Using Commercially Available Sonars," Technical Report REMR-CO-11, U.S. Army Engineer Waterways Experiment Station, Vicksburg, MS.

Langelier 1936
Langelier, W. F. 1936 (Oct). "The Analytical Control of Anti-Corrosion Water Treatment," *Journal of the American Water Works Association,* Vol 28, No. 10, pp 1500-1521.

Lanigan 1992
Lanigan, Carl, A. 1992. "Continuous Deformation Monitoring System (CDMS)," Technical Report REMR-CS-39, U.S. Army Engineer Waterways Experiment Station, Vicksburg, MS.

Lauer 1956
Lauer, K. R., and Slate, F. O. 1956 (Jun). "Autogenous Healing of Cement Paste," *ACI Journal,* Proceedings, Vol 27, No. 10, pp 1083-1098.

Liu 1980
Liu, T. C. 1980 (Jul). "Maintenance and Preservation of Concrete Structures; Report 3, Abrasion-Erosion Resistance of Concrete," Technical Report C-78-4, U.S. Army Engineer Waterways Experiment Station, Vicksburg, MS.

Liu and Holland 1981
Liu, T. C., and Holland, T. C. 1981 (Mar). "Design of Dowels for Anchoring Replacement Concrete to Vertical Lock Walls," Technical Report SL-81-1, U.S. Army Engineer Waterways Experiment Station, Vicksburg, MS.

Liu and McDonald 1981
Liu, T. C., and McDonald, J. E. 1981. "Abrasion-Erosion Resistance of Fiber-Reinforced Concrete," *Cement, Concrete, and Aggregates,* Vol 3, No. 2.

Mailvaganam 1992
Mailvaganam, Noel P. 1992. "Repair and Protection of Concrete Structures," CRC Press, London.

Malhotra 1976
Malhotra, V. M. 1976. "Testing Hardened Concrete: Nondestructive Methods," ACI Monograph No. 9, Detroit, MI.

Marold 1992
Marold, W. J. 1992 (Feb). "Design of the Boney Falls RCC Emergency Spillway," *Roller Compacted Concrete III,* American Society of Civil Engineers, pp 476-490.

McDonald 1980
McDonald, J. E. 1980 (Apr). "Maintenance and Preservation of Concrete Structures; Report 2, Repair of Erosion-Damaged Structures," Technical Report C-78-4, U.S. Army Engineer Waterways Experiment Station, Vicksburg, MS.

McDonald 1986
McDonald, James E. 1986 (Nov). "Repair of Waterstop Failures: Case Histories," Technical Report REMR-CS-4, U.S. Army Engineer Waterways Experiment Station, Vicksburg, MS.

McDonald 1987a
McDonald, James E. 1987 (Jul). "Precast Concrete Stay-in-Place Forming System for Lock Wall Rehabilitation," *The REMR Bulletin,* Vol 4, No. 1, U.S. Army Engineer Waterways Experiment Station, Vicksburg, MS.

McDonald 1987b
McDonald, J. E. 1987b (Dec). "Rehabilitation of Navigation Lock Walls: Case Histories," Technical Report REMR-CS-13, U.S. Army Engineer Waterways Experiment Station, Vicksburg, MS.

McDonald 1988
McDonald, J. E. 1988 (Jul). "A Precast Stay-in-Place Forming System for Lock Wall Rehabilitation," Video Report REMR-CS-1, U.S. Army Engineer Waterways Experiment Station, Vicksburg, MS.

McDonald 1989
McDonald, J. E. 1989 (Feb). "Evaluation of Vinylester Resin for Anchor Embedment in Concrete," Technical Report REMR-CS-20, U.S. Army Engineer Waterways Experiment Station, Vicksburg, MS.

McDonald 1990
McDonald, J. E. 1990 (Oct). "Anchor Embedment in Hardened Concrete Under Submerged Conditions," Technical Report REMR-CS-33, U.S. Army Engineer Waterways Experiment Station, Vicksburg, MS.

McDonald 1991
McDonald, James E. 1991 (Mar). "Properties of Silica-Fume Concrete," Technical Report REMR-CS-32, U.S. Army Engineer Waterways Experiment Station, Vicksburg, MS.

McDonald and Curtis 1995
McDonald, J. E, and Curtis, N. 1995 (Apr). "Applications of Precast Concrete in Repair of Civil Works Structures." Technical Report (REMR-CS-48), U.S. Army Engineer Waterways Experiment Station, Vicksburg, MS.

Mech 1989
Mech, George J. 1989 (Oct). "Rehabilitation of Peoria Lock Using Preplaced-Aggregate Concrete," *The REMR Bulletin*, Vol 6, No. 4, U.S. Army Engineer Waterways Experiment Station, Vicksburg, MS.

Meyers 1994
Meyers, John G. 1994 (Aug/Sep). "Slabjacking Sunken Concrete," *Concrete Repair Digest*, Vol 6, No. 4, The Aberdeen Group, Addison, Il.

Miles 1993
Miles, William R. 1993 (Oct). "Comparison of Cast-in-Place Concrete Versus Precast Concrete Stay-in-Place Forming Systems for Lock Wall Rehabilitation," Technical Report REMR-CS-41, U.S. Army Engineer Waterways Experiment Station, Vicksburg, MS.

Minnotte 1952
Minnotte, J. S. 1952 (Oct). "Lock No. 5 Monongahela River Refaced by Grout Intrusion Method," *Civil Engineering*, Vol 22, pp 872-875.

Mitscher 1992
Mitscher, Kurt A. 1992 (Dec). "Sheet-Pile and Precast Concrete U-Flume Low-Flow Channels for the Blue River Paved Reach Project," 1991 Corps of Engineer Structural Engineering Conference, pp 471-480.

Monari 1984
Monari, F. 1984. "Waterproof Covering for the Upstream of the Lago Nero Dam," *Proceedings, International Conference on Geomembranes*, Denver, CO, pp 105-110.

Monari and Scuero 1991
Monari, F., and Scuero, A. M. 1991. "Aging of Concrete Dams: The Use of Geocomposites for Repair and Future Protection," International Commission on Large Dams, 17th Congress, June 17-21.

Montani 1993
Montani, Rick. 1993 (May/Jun). "High Molecular Weight Methacrylates," *Concrete Repair Bulletin*, Vol 6, No. 3, pp 6-9.

Morang 1987
Morang, A. 1987 (Dec). "Side-Scan Sonar Investigation of Breakwaters at Calumet and Burns Harbors on Southern Lake Michigan," Miscellaneous Paper CERC-87-20, U.S. Army Engineer Waterways Experiment Station, Vicksburg, MS.

Morey 1974
Morey, R. M. 1974 (Mar). "Application of Downward Looking Impulse Radar," *Proceedings of 13th Annual Canadian Hydrographic Conference*, Canada Center for Inland Waters, Burlington, Ontario.

NACE International 1991
National Association of Corrosion Engineers. 1991 (Oct). "Coatings for Concrete Surfaces in Non-Immersion and Atmospheric Services," Standard Recommended Practice RP0591-91, Houston,TX.

Neeley 1988
Neeley, B. D. 1988 (Apr). "Evaluation of Concrete Mixtures for Use in Underwater Repairs," Technical Report REMR-CS-18, U.S. Army Engineer Waterways Experiment Station, Vicksburg, MS.

Neeley and Wickersham 1989
Neeley, B. D., and Wickersham, J. 1989 (Oct). "Repair of Red Rock Dam," *Concrete International: Design and Construction*, Vol 11, No. 10, American Concrete Institute, Detroit, MI.

Neeley, Saucier, and Thornton 1990
Neeley, B. D., Saucier, K. L., and Thornton, H. T., Jr. 1990 (Nov). "Laboratory Evaluation of Concrete Mixtures and Techniques for Underwater Repairs," Technical

Report REMR-CS-34, U.S. Army Engineer Waterways Experiment Station, Vicksburg, MS.

Norman, Campbell, and Garner 1988
Norman, C. D., Campbell, R. L., Jr., and Garner, S. 1988 (Aug). "Analysis of Concrete Cracking in Lock Wall Resurfacing," Technical Report REMR-CS-15, U.S. Army Engineer Waterways Experiment Station, Vicksburg, MS.

Oswalt 1971
Oswalt, N. R. 1971. "Pomona Dam Outlet Stilling Basin Modifications," Memorandum Report, U.S. Army Engineer Waterways Experiment Station, Vicksburg, MS.

PCA 1990
"Upper Michigan Dam Rehabilitated with RCC." 1990 (Spring/Summer). *RCC Newsletter*, Vol 5, No. 1, Portland Cement Association, Skokie, IL.

Pfeifer and Scali 1981
Pfeifer, D. W., and Scali, M. J. 1981 (Dec). "Concrete Sealers for Protection of Bridge Structures," National Cooperative Highway Research Program Report No. 244, Transportation Research Board, Washington, DC.

Pinney 1991
Pinney, Stephen G. 1991 (Feb). "Preparation, Application and Inspection of Coatings for Concrete," *REMR Bulletin*, Vol 8, No. 1, Vicksburg, MS.

Popovics and McDonald 1989
Popovics, S., and McDonald, W. E. 1989 (Apr). "Inspection of the Engineering Condition of Underwater Concrete Structures," Technical Report REMR-CS-9, U.S. Army Engineer Waterways Experiment Station, Vicksburg, MS.

Rail and Haynes 1991
Rail, R. D., and Haynes, H, H. 1991 (Dec). "Underwater Stilling Basin Repair Techniques Using Precast or Prefabricated Elements," Technical Report REMR-CS-38, U.S. Army Engineer Waterways Experiment Station, Vicksburg, MS.

Ramakrishnan 1992
Ramakrishnan, V. 1992 (Aug). "Latex-Modified Concretes and Mortars," National Cooperative Highway Research Program, Synthesis of Highway Practice 179, Transportation Research Board, Washington, DC.

Schrader 1980
Schrader, E. K. 1980 (Oct). "Repair of Waterstop Failures," *ASCE Journal of the Energy Division*, Vol 106, No. EY2, pp 155-163.

Schrader 1981
Schrader, E. K. 1981 (Jun). "Deterioration and Repair of Concrete in the Lower Monumental Navigation Lock Wall," Miscellaneous Paper SL-81-9, U.S. Army Engineer Waterways Experiment Station, Vicksburg, MS.

Schrader 1992
Schrader, Ernest K. 1992 (Dec). "Mistakes, Misconceptions, and Controversial Issues Concerning Concrete and Concrete Repairs," *Concrete International*, American Concrete Institute, Vol 14, No. 11, pp 54-59.

Schrader and Kaden 1976
Schrader, E. K., and Kaden, R. A. 1976 (Jul-Aug). "Stilling Basin Repairs at Dworshak Dam," *The Military Engineer*, Vol 68, No. 444, Alexandria, VA.

Simons 1992
Simons, B. P. 1992 (Mar). "Abrasion Testing for Suspended Sediment Loads," *Concrete International*, Vol 14, No. 3, pp.

Smith 1987
Smith, A. P. 1987 (Apr). "New Tools and Techniques for the Underwater Inspection of Waterfront Structures, OTC 5390, *19th Annual Offshore Technology Conference*, Houston, TX.

Solomon and Jaques 1994
Solomon, Joseph, and Jaques, Mike. 1994 (Jun-Jul). "Stopping Leaks With Polyurethane Grouts," *Concrete Repair Digest*, Vol 5, No. 3, pp 180-185.

SONEX 1983
SONEX, LTD. 1983 (Sep). "Sonic Inspection of Ice Harbor Dam Spillway Stilling Basin," prepared for U.S. Army Corps of Engineers, U.S. Army Engineer District, Walla Walla, Walla Walla, WA, under Contract DACW39-83-M-3397, revised Feb 1984.

SONEX 1984
SONEX, LTD. 1984 (Jan). "Final Report: High Resolution Acoustic Survey, Folsom Dam Stilling Basin Floor," prepared for U.S. Army Engineer Waterways Experiment

Station, Vicksburg, MS, under Purchase Order DACW39-83-M-4340.

Stowe and Thornton 1984
Stowe, R. L., and Thornton, H. T., Jr. 1984 (Sep). "Engineering Condition Survey of Concrete in Service," Technical Report REMR-CS-1, U.S. Army Engineer Waterways Experiment Station, Vicksburg, MS.

Stowe and Campbell 1989
Stowe, R. L., and Campbell, R. L., Sr. 1989. "User's Guide: Maintenance and Repair Materials Database for Concrete and Steel Structures," Technical Report REMR-CS-27, U.S. Army Corps of Engineers, Waterways Experiment Station, Vicksburg, MS.

Stratton, Alexander, and Nolting 1982
Stratton, F. W., Alexander, R., and Nolting, W. 1982 (May). "Development and Implementation of Concrete Girder Repair by Post-Reinforcement," Report No. FHWA-KS-82-1, Kansas Department of Transportation, Topeka, 31 pp.

Sumner 1993
Sumner, Andrew C. 1993. "Rehabilitation of Crescent and Vischer Ferry Dams, Construction Techniques & Problem Solutions," *Geotechnical Practice in Dam Rehabilitation*, American Society of Civil Engineers, New York, NY.

Thornton 1985
Thornton, H. T., Jr. 1985 (Mar). "Corps-BuRec Effort Results in High-Resolution Acoustic Mapping System," *The REMR Bulletin*, Vol 2, No. 1, U.S. Army Engineer Waterways Experiment Station, Vicksburg, MS.

Thornton and Alexander 1987
Thornton, H. T., and Alexander, A. M. 1987 (Dec). "Development of Nondestructive Testing Systems for In Situ Evaluation of Concrete Structures," Technical Report REMR-CS-10, U.S. Army Engineer Waterways Experiment Station, Vicksburg, MS.

Thornton and Alexander 1988
Thornton, H. T., Jr., and Alexander, A. M. 1988 (Mar). "Ultrasonic Pulse-Echo Measurements of the Concrete Sea Wall at Marina Del Rey, Los Angeles County, California," *The REMR Bulletin*, Vol 5, No. 1, U.S. Army Engineer Waterways Experiment Station, Vicksburg, MS.

Trout 1994
Trout, John. 1994 (Mar-Apr). "Comparing High Pressure and Low Pressure Injection," *Concrete Repair Bulletin*, Vol 7, No. 2, pp 20-21.

U.S. Army Engineer District, Walla Walla 1979
U.S. Army Engineer District, Walla Walla. 1979 (Jan). "Dworshak Dam and Reservoir, Inspection Report No. 6," Walla Walla, WA.

U.S. Army Engineer Division, Missouri River 1974
U.S. Army Engineer Division, Missouri River. 1974 (Apr). "Development of Equipment and Techniques for Pneumatic Application of Portland Cement Mortar in Shallow Patches," Omaha, NE.

U.S. Department of Transportation 1989
U.S. Department of Transportation. 1989 (Nov). "Underwater Inspection of Bridges," Report No. FHWA-DP-80-1, Federal Highway Administration, Washington, DC.

Warner 1984
Warner, J. 1984 (Oct). "Selecting Repair Materials," *Concrete Construction*, Vol 29, No. 10, pp 865-871.

Webster and Kukacka 1988
Webster, R. P., and Kukacka, L. E. 1988 (Jan). "In Situ Repair of Deteriorated Concrete in Hydraulic Structures: Laboratory Study," Technical Report REMR-CS-11, U.S. Army Engineer Waterways Experiment Station, Vicksburg, MS.

Webster, Kakacka, and Elling 1989
Webster, R. P., Kukacka, L. E., and Elling, D. 1989 (Apr). "In Situ Repair of Deteriorated Concrete in Hydraulic Structures: A Field Study," Technical Report REMR-CS-21, U.S. Army Engineer Waterways Experiment Station, Vicksburg, MS.

Webster, Kakacka, and Elling 1990
Webster, R. P., Kukacka, L. E., and Elling, D. 1990 (Sep). "In Situ Repair of Deteriorated Concrete in Hydraulic Structures: Epoxy Injection Repair of a Bridge Pier," Technical Report REMR-CS-30, U.S. Army Engineer Waterways Experiment Station, Vicksburg, MS.

Wickersham 1987
Wickersham, J. 1987 (Dec). "Concrete Rehabilitation at Lock and Dam No. 20, Mississippi River," *The REMR*

Bulletin, Vol 4, No. 4, U.S. Army Engineer Waterways Experiment Station, Vicksburg, MS.

Wong, Feldsher, Wright, and Johnson 1992
Wong, Noel C., Feldsher, Theodore B., Wright, Robert S., and Johnson, David H. 1992 (Feb). "Final Design and Construction of Gibraltar Dam Strengthening," *Roller Compacted Concrete III*, American Society of Civil Engineers, pp 440-458.

Wong, Forrest, and Lo 1993
Wong, Noel C., Forrest, Michael P., and Lo, Sze-Hang. 1993 (Sep). "Littlerock Dam Rehabilitation: Another RCC Innovation," *Proceedings from the 10th ASDSO Annual Conference*, Association of State Dam Safety Officials, pp 303-314.

Xanthakos 1979
Xanthakos, P. P. 1979. *Slurry Walls*, McGraw-Hill Book Co., New York.

Appendix B
Glossary

Terms related to evaluation and repair of concrete structures as used herein are defined as follows:

Abrasion resistance
Ability of a surface to resist being worn away by rubbing and friction.

Acrylic resin
One of a group of thermoplastic resins formed by polymerizing the esters or amides of acrylic acid; used in concrete construction as a bonding agent or surface sealer.

Adhesives
The group of materials used to join or bond similar or dissimilar materials; for example, in concrete work, the epoxy resins.

Air-water jet
A high-velocity jet of air and water mixed at the nozzle; used in cleanup of surfaces of rock or concrete such as horizontal construction joints.

Alkali-aggregate reaction
Chemical reaction in mortar or concrete between alkalies (sodium and potassium) from portland cement or other sources and certain constituents of some aggregates; under certain conditions, deleterious expansion of the concrete or mortar may result.

Alkali-carbonate rock reaction
The reaction between the alkalies (sodium and potassium) in portland cement and certain carbonate rocks, particularly calcitic dolomite and dolomitic limestones, present in some aggregates; the products of the reaction may cause abnormal expansion and cracking of concrete in service.

Alkali reactivity (of aggregate)
Susceptibility of aggregate to alkali-aggregate reaction.

Alkali-silica reaction
The reaction between the alkalies (sodium and potassium) in portland cement and certain siliceous rocks or minerals, such as opaline chert and acidic volcanic glass, present in some aggregates; the products of the reaction may cause abnormal expansion and cracking of concrete in service.

Autogenous healing
A natural process of closing and filling of cracks in concrete or mortar when kept damp.

Bacterial corrosion
The destruction of a material by chemical processes brought about by the activity of certain bacteria which may produce substances such as hydrogen sulfide, ammonia, and sulfuric acid.

Blistering
The irregular raising of a thin layer at the surface of placed mortar or concrete during or soon after completion of the finishing operation, or in the case of pipe after spinning; also bulging of the finish plaster coat as it separates and draws away from the base coat.

Bug holes
Small regular or irregular cavities, usually not exceeding 15 mm in diam, resulting from entrapment of air bubbles in the surface of formed concrete during placement and compaction.

Butyl stearate
A colorless oleaginous, practically odorless material ($C_{17}H_{35}COOC_4H_9$) used as an admixture for concrete to provide dampproofing.

Cavitation damage
Pitting of concrete caused by implosion; i.e., the collapse of vapor bubbles in flowing water which form in areas of low pressure and collapse as they enter areas of higher pressure.

Chalking
Formation of a loose powder resulting from the disintegration of the surface of concrete or an applied coating such as cement paint.

Checking
Development of shallow cracks at closely spaced, but irregular, intervals on the surface of plaster, cement paste, mortar, or concrete.

Cold-joint lines
Visible lines on the surfaces of formed concrete indicating the presence of joints where one layer of concrete had hardened before subsequent concrete was placed.

Concrete, preplaced-aggregate
Concrete produced by placing coarse aggregate in a form and later injecting a portland-cement-sand grout, usually with admixtures, to fill the voids.

Corrosion
Destruction of metal by chemical, electrochemical, or electrolytic reaction with its environment.

Cracks, active*
Those cracks for which the mechanism causing the cracking is still at work. Any crack that is still moving.

Cracks, dormant*
Those cracks not currently moving or which the movement is of such magnitude that the repair material will not be affected.

Craze cracks
Fine, random cracks or fissures in a surface of plaster, cement paste, mortar, or concrete.

Crazing
The development of craze cracks; the pattern of craze cracks existing in a surface. (See also Checking.)

Dampproofing
Treatment of concrete or mortar to retard the passage or absorption of water or water vapor, either by application of a suitable coating to exposed surfaces or by use of a suitable admixture, treated cement, or preformed films such as polyethylene sheets under slabs on grade. (See also Vapor barrier.)

D-cracking
A series of cracks in concrete near and roughly parallel to joints, edges, and structural cracks.

Delamination
A separation along a plane parallel to a surface as in the separation of a coating from a substrate or the layers of a coating from each other, or in the case of a concrete slab, a horizontal splitting, cracking, or separation of a slab in a plane roughly parallel to, and generally near, the upper surface; found most frequently in bridge decks and caused by the corrosion of reinforcing steel or freezing and thawing; similar to spalling, scaling, or peeling except that delamination affects large areas and can often be detected only by tapping.

---

* All definitions are in accordance with ACI 116R except those denoted by an asterisk.

Deterioration
Decomposition of material during testing or exposure to service. (See also Disintegration.)

Diagonal crack
In a flexural member, an inclined crack caused by shear stress, usually at about 45 deg to the neutral axis of a concrete member; a crack in a slab, not parallel to the lateral or longitudinal directions.

Discoloration
Departure of color from that which is normal or desired.

Disintegration
Reduction into small fragments and subsequently into particles.

Dry-mix shotcrete
Shotcrete in which most of the mixing water is added at the nozzle.

Drypacking
Placing of zero slump, or near zero slump, concrete, mortar, or grout by ramming it into a confined space.

Durability
The ability of concrete to resist weathering action, chemical attack, abrasion, and other conditions of service.

Dusting
The development of a powdered material at the surface of hardened concrete.

Efflorescence
A deposit of salts, usually white, formed on a surface, the substance having emerged in solution from within concrete or masonry and subsequently having been precipitated by evaporation.

Epoxy concrete
A mixture of epoxy resin, catalyst, fine aggregate, and coarse aggregate. (See also Epoxy mortar, Epoxy resin, and Polymer concrete.)

Epoxy mortar
A mixture of epoxy resin, catalyst, and fine aggregate. (See also Epoxy resin.)

Epoxy resin
A class of organic chemical bonding systems used in the preparation of special coatings or adhesives for concrete or as binders in epoxy resin mortars and concretes.

Erosion
Progressive disintegration of a solid by the abrasive or cavitation action of gases, fluids, or solids in motion. (See also Abrasion resistance and Cavitation damage.)

Ettringite
A mineral, high-sulfate calcium sulfoaluminate $(3CaO \cdot Al_2O_3 \cdot 3CaSO_4 \cdot 32H_2O)$ also written as $Ca_6[Al(OH)_6]_2 \cdot 24H_2O[(SO_4)3 \cdot (1\text{-}1/2) \ H_2O]$ occurring in nature or formed by sulfate attack on mortar and concrete; the product of the principal expansion-producing reaction in expansive cements; designated as "cement bacillus" in older literature.

Evaluation*
Determining the condition, degree of damage or deterioration, or serviceability and, when appropriate, indicating the need for repair, maintenance, or rehabilitation. (See also Repair, Maintenance, and Rehabilitation.)

Exfoliation
Disintegration occurring by peeling off in successive layers; swelling up and opening into leaves or plates like a partly opened book.

Exudation
A liquid or viscous gel-like material discharge through a pore, crack, or opening in the surface of concrete.

Feather edge
Edge of a concrete or mortar patch or topping that is beveled at an acute angle.

Groove joint
A joint created by forming a groove in the surface of a pavement, floor slab, or wall to control random cracking.

Hairline cracks
Cracks in an exposed concrete surface having widths so small as to be barely preceptible.

Honeycomb
Voids left in concrete due to failure of the mortar to effectively fill the spaces among coarse aggregate particles.

Incrustation
A crust or coating, generally hard, formed on the surface

---

* All definitions are in accordance with ACI 116R except those denoted by an asterisk.

of concrete or masonry construction or on aggregate particles.

Joint filler
Compressible material used to fill a joint to prevent the infiltration of debris and to provide support for sealants.

Joint sealant
Compressible material used to exclude water and solid foreign material from joints.

Laitance
A layer of weak and nondurable material containing cement and fines from aggregates, brought by bleeding water to the top of overwet concrete, the amount of which is generally increased by overworking or over-manipulating concrete at the surface by improper finishing or by job traffic.

Latex
A water emulsion of a high molecular-weight polymer used especially in coatings, adhesives, and leveling and patching compounds.

Maintenance*
Taking periodic actions that will prevent or delay damage or deterioration or both. (See also Repair.)

Map cracking
See Crazing.

Microcracks
Microscopic cracks within concrete.

Monomer
An organic liquid of relatively low molecular weight that creates a solid polymer by reacting with itself or other compounds of low molecular weight or with both.

Overlay
A layer of concrete or mortar, seldom thinner than 25 mm (1 in.), placed on and usually bonded onto the worn or cracked surface of a concrete slab to restore or improve the function of the previous surface.

Pattern cracking
Intersecting cracks that extend below the surface of hardened concrete; caused by shrinkage of the drying surface which is restrained by concrete at greater depth where little or no shrinkage occurs; vary in width and depth from fine and barely visible to open and well defined.

Peeling
A process in which thin flakes of mortar are broken away from a concrete surface, such as by deterioration or by adherence of surface mortar to forms as forms are removed.

Pitting
Development of relatively small cavities in a surface caused by phenomena such as corrosion or cavitation, or in concrete localized disintegration such as a popout.

Plastic cracking
Cracking that occurs in the surface of fresh concrete soon after it is placed and while it is still plastic.

Plastic shrinkage cracks
See Plastic cracking.

Polyester
One of a large group of synthetic resins, mainly produced by reaction of dibasic acids with dihydroxy alcohols, commonly prepared for application by mixing with a vinyl-group monomer and free-radical catalyst at ambient temperatures and used as binders for resin mortars and concretes, fiber laminates (mainly glass), adhesives, and the like. (See also Polymer concrete.)

Polyethylene
A thermoplastic high-molecular-weight organic compound used in formulating protective coatings; in sheet form, used as a protective cover for concrete surfaces during the curing period, or to provide a temporary enclosure for construction operations.

Polymer
The product of polymerization; more commonly, a rubber or resin consisting of large molecules formed by polymerization.

Polymer concrete
Concrete in which an organic polymer serves as the binder; also known as resin concrete; sometimes erroneously employed to designate hydraulic-cement mortars or concretes in which part or all of the mixing water is replaced by an aqueous dispersion of a thermoplastic copolymer.

Polymer-cement concrete
A mixture of water, hydraulic cement, aggregate, and a monomer or polymer polymerized in place when a monomer is used.

Polymerization
The reaction in which two or more molecules of the same substance combine to form a compound containing the same elements in the same proportions, but of higher molecular weight, from which the original substance can be generated, in some cases only with extreme difficulty.

Polystyrene resin
Synthetic resins varying in color from colorless to yellow formed by the polymerization of styrene, or heated, with or without catalysts, that may be used in paints for concrete or for making sculptured molds or as insulation.

Polysulfide coating
A protective coating system prepared by polymerizing a chlorinated alkylpolyether with an inorganic polysulfide.

Polyurethane
Reaction product of an isocyanate with any of a wide variety of other compounds containing an active hydrogen group; used to formulate tough, abrasion-resistant coatings.

Polyvinyl acetate
Colorless, permanently thermoplastic resin, usually supplied as an emulsion or water-dispersible powder characterized by flexibility, stability toward light, transparency to ultraviolet rays, high dielectric strength, toughness, and hardness; the higher the degree of polymerization, the higher the softening temperature; may be used in paints for concrete.

Polyvinyl chloride
A synthetic resin prepared by the polymerization of vinyl chloride; used in the manufacture of nonmetallic waterstops for concrete.

Popout
The breaking away of small portions of concrete surface due to internal pressure, which leaves a shallow, typically conical, depression.

Pot life
Time interval after preparation during which a liquid or plastic mixture is usable.

Reactive aggregate
Aggregate containing substances capable of reacting chemically with the products of solution or hydration of the portland cement in concrete or mortar under ordinary

conditions of exposure, resulting in some cases in harmful expansion, cracking, or staining.

Rebound hammer
An apparatus that provides a rapid indication of the mechanical properties of concrete based on the distance of rebound of a spring-driven missile.

Rehabilitation
The process of repairing or modifying a structure to a desired useful condition.

Repair
Replace or correct deteriorated, damaged, or faulty materials, components, or elements of a structure.

Resin
A natural or synthetic, solid or semisolid organic material of indefinite and often high molecular weight having a tendency to flow under stress that usually has a softening or melting range and usually fractures conchoidally.

Resin mortar (or concrete)
See Polymer concrete.

Restraint (of concrete)
Restriction of free movement of fresh or hardened concrete following completion of placement in formwork or molds or within an otherwise confined space; restraint can be internal or external and may act in one or more directions.

Rock pocket
A porous, mortar-deficient portion of hardened concrete consisting primarily of coarse aggregate and open voids, caused by leakage of mortar from form, separation (segregation) during placement, or insufficient consolidation. (See also Honeycombing.)

Sandblasting
A system of cutting or abrading a surface such as concrete by a stream of sand ejected from a nozzle at high speed by compressed air; often used for cleanup of horizontal construction joints or for exposure of aggregate in architectural concrete.

Sand streak
A streak of exposed fine aggregate in the surface of formed concrete that is caused by bleeding.

Scaling
Local flaking or peeling away of the near-surface portion of hardened concrete or mortar; also of a layer from metal. (See also Peeling and Spalling.) (Note: Light scaling of concrete does not expose coarse aggregate; medium scaling involves loss of surface mortar to 5 to 10 mm in depth and exposure of coarse aggregate; severe scaling involves loss of surface mortar to 5 to 10 mm in depth with some loss of mortar surrounding aggregate particles 10 to 20 mm in depth; very severe scaling involves loss of coarse-aggregate particles as well as mortar generally to a depth greater than 20 mm.)

Shotcrete
Mortar or concrete pneumatically projected at high velocity onto a surface; also known as air-blown mortar; also pneumatically applied mortar or concrete, sprayed mortar, and gunned concrete. (See also Dry-mix shotcrete and Wet-mix shotcrete.)

Shrinkage
Volume decrease caused by drying and chemical changes; a function of time but not temperature or of stress caused by external load.

Shrinkage crack
Crack due to restraint of shrinkage.

Shrinkage cracking
Cracking of a structure or member from failure in tension caused by external or internal restraints as reduction in moisture content develops or as carbonation occurs, or both.

Spall
A fragment, usually in the shape of a flake, detached from a larger mass by a blow, action of weather, pressure, or expansion within the larger mass; a small spall involves a roughly circular depression not greater than 20 mm in depth nor 150 mm in any dimension; a large spall may be roughly circular or oval or, in some cases, elongated more than 20 mm in depth and 150 mm in greatest dimension.

Stalactite
A downward-pointing deposit formed as an accretion of mineral matter produced by evaporation of dripping water from the surface of concrete, commonly shaped like an icicle.

Stalagmite
An upward-pointing deposit formed as an accretion of mineral matter produced by evaporation of dripping water, projecting from the surface of concrete, and commonly conical in shape.

Spalling
The development of spalls.

Sulfate attack
Chemical or physical reaction, or both, between sulfates, usually in soil or ground water and concrete or mortar, primarily with calcium aluminate hydrates in the cement-paste matrix, often causing deterioration.

Sulfate resistance
Ability of concrete or mortar to withstand sulfate attack. (See also Sulfate attack.)

Swiss hammer
See Rebound hammer.

Temperature cracking
Cracking as a result of tensile failure caused by temperature drop in members subjected to external restraints or temperature differential in members subjected to internal restraints.

Thermal shock
The subjection of newly hardened concrete to a rapid change in temperature which may be expected to have a potentially deleterious effect.

Thermoplastic
Becoming soft when heated and hard when cooled.

Thermosetting
Becoming rigid by chemical reaction and not remeltable.

Transverse cracks
Cracks that develop at right angles to the long direction of a member.

Tremie
A pipe or tube through which concrete is deposited underwater, having at its upper end a hopper for filling and a bail for moving the assemblage.

Tremie concrete
Subaqueous concrete placed by means of a tremie.

Tremie seal
The depth to which the discharge end of the tremie pipe is kept embedded in the fresh concrete that is being placed; a layer of tremie concrete placed in a cofferdam for the purpose of preventing the intrusion of water when the cofferdam is dewatered.

Vapor barrier
A membrane placed under concrete floor slabs that are placed on grade and intended to retard transmission of water vapor.

Waterstop
A thin sheet of metal, rubber, plastic, or other material inserted across a joint to obstruct seepage of water through the joint.

Water void
Void along the underside of an aggregate particle or reinforcing steel which formed during the bleeding period and initially filled with bleed water.

Weathering
Changes in color, texture, strength, chemical composition, or other properties of a natural or artificial material caused by the action of the weather.

Wet-mix shotcrete
Shotcrete in which the ingredients, including mixing water, are mixed before introduction into the delivery hose; accelerator if used, is normally added at the nozzle.

Appendix C
Abbreviations

ACI	American Concrete Institute
ASTM	American Society for Testing and Materials
AWA	antiwashout admixture
CDMS	Continuous Deformation Monitoring System
CE	Corps of Engineers
CERC	Coastal Engineering Research Center
CEWES-SC	U.S. Army Engineer Waterways Experiment Station, Structures Laboratory, Concrete Technology Division
CMU	concrete masonry units
CRD	Concrete Research Division, Handbook for Concrete and Cement
EM	Engineer Manual
EP	Engineer Pamplet
ER	Engineer Regulation
FHWA	Federal Highway Administration
GPS	Global Positioning System
HAC	high alumina cement
HMWM	high molecular weight methacrylate
HQUSACE	Headquarters, U.S. Army Corps of Engineers
HRWRA	high-range water-reducing admixture
ICOLD	International Commission on Large Dams
MPC	magnesium phosphate cement
MSA	maximum size aggregate
MSDS	Manufacturer's Safety Data Sheet
NACE	National Association of Corrosion Engineers
NAVSTAR	Navigation Satellite Timing and Ranging
NCHRP	National Cooperative Highway Research Program
NDT	nondestructive testing
OCE	Office, Chief of Engineers
PC	polymer concrete
PCA	Portland Cement Association
PIC	polymer-impregnated concrete
PPCC	polymer portland-cement concrete
PMF	Probable Maximum Flood
PVC	polyvinyl chloride
R-values	rebound readings
RCC	roller-compacted concrete
REMR	Repair, Evaluation, Maintenance, and Rehabilitation Research Program
ROV	remotely operated vehicle
TM	Technical Manual
TOA	time of arrival
UPE	ultrasonic pulse-echo
UV	ultraviolet
w/c	water-cement ratio
WES	Waterways Experiment Station
WRA	water-reducing admixture

LaVergne, TN USA
18 December 2009
167498LV00001B/9/A

This manual provides guidance on evaluating the condition of the concrete in a structure, relating the condition of the concrete to the underlying cause or causes of that condition, selecting an appropriate repair material and method for any deficiency found, and using the selected materials and methods to repair or rehabilitate the structure. Guidance is also included on maintenance of concrete and on preparation of concrete investigation reports for repair and rehabilitation projects. Considerations for certain specialized types of rehabilitation projects are also given.

ISBN 1-4101-0743-4

9 781410 107435